D0349801

THE CENTRAL SCIENCE

THE CENTRAL SCIENCE

Essays on the Uses of Chemistry

Edited by

GEORGE B. KAUFFMAN
Professor of Chemistry
California State University, Fresno
Fresno, California

and

H. HARRY SZMANT
Chairman, Department of Chemistry
and Chemical Engineering
University of Detroit
Detroit, Michigan

TEXAS CHRISTIAN UNIVERSITY PRESS • FORT WORTH

Library of Congress Cataloging in Publication Data
Main entry under title:
The Central science
Updated papers of the "Multifaceted contributions to chemistry" workshop organized for the 175th National Meeting of the American Chemical Society, sponsored by the ACS Council Committee for Public Relations and the ACS Council Committee for Chemical Education, and held in March 1978, at Anaheim, Calif.
I. Chemistry — Congresses. I. Kauffman, George B., 1930-
II. Szmant, H. Harry (Herman Harry), 1918-
III. American Chemical Society. Meeting (175th:1978:Anaheim, Calif.) IV. ACS Council Committee for Public Relations. V. ACS Council Committee for Chemical Education.
QD1.C44 1984 540 83-18054
ISBN 0-912646-84-5

Second printing, 1985

Designed by Janet Brooks
Whitehead & Whitehead / Austin

Printed and bound in the United States

TABLE OF CONTENTS

PREFACE

The objective of the "Multifaceted Contributions to Chemistry" workshop symposium, organized for the 175th National Meeting of the American Chemical Society in March, 1978 at Anaheim, California, was to bring together a dozen recognized authorities from academic, industrial, and governmental institutions in order to provide an overview of the multifaceted contributions that chemistry makes to our individual and societal well-being. The symposium, sponsored by the ACS Council Committee for Public Relations and the Council Committee for Chemical Education, was designed to publicize the role chemistry plays in almost every aspect of our civilization. This book consists of the updated contributions of the symposium participants. It is intended to be particularly attractive to an audience of high school students, their teachers, counselors, librarians, PTA board members, and parents.

In recent years it has become commonplace to refer to chemistry as "the central science." We are indebted to Prentice-Hall, Inc., Englewood Cliffs, New Jersey, for permission to use this phrase, adopted from the popular text, "Chemistry: The Central Science" (1977, 1981) by Theodore L. Brown and H. Eugene LeMay. Those of us who have a professional background in chemical science or technology do not have any difficulty in recognizing the fact that chemistry, among all of the branches of the natural sciences, is singled out for such a distinction. Most lay persons should also recognize the special role of chemical science and technology in connection with the understanding of the nature and the interesting and useful transformations of matter from one kind to another.

The spectrum of topics dealt with in this volume ranges from the understanding of the most grandiose and mysterious of all systems, namely the cosmos and the beginnings of all life processes, to the understanding of the most complex but compact system, namely that of the human being and his behavioral patterns. It also touches upon such diverse topics as the understanding of the oceans; prehistoric and the most recent events in human affairs; the solutions to our problems of the supply and delivery of food, energy, and health care; the reinforcement of our criminal justice system; the workings of our economy; warfare and its "spin-offs"; and the development of materials and techniques for the satisfaction of both materialistic and nonmaterialistic needs in the form of consumer goods and objects of art, respectively.

We hope that the format and content of this volume will be an effective means of awakening an interest in and promoting an appreciation of the multifaceted positive contributions that chemical science and technology make to the quality of our daily lives.

I wish to acknowledge the relentless and successful search by my co-editor for a publisher for this book. Also, I thank all of the contributors for their patience and understanding in our attempts to mold the collection of essays into an effective educational tool, and Ms. Judy Alter, editor of the Texas Christian University Press, for her skillful efforts to bring an acceptable degree of uniformity to the volume.

H. Harry Szmant
Professor of Chemistry
University of Detroit

INTRODUCTION

Chemistry has been characterized as the most utilitarian of all the experimental sciences. Its interdisciplinary position between physics and biology makes possible many practical applications to the welfare of mankind. It overlaps and permeates both of these sister sciences to such an extent that there is, in fact, no clear distinction among them; physical chemistry fades into chemical and nuclear physics just as biochemistry and medicinal chemistry in turn mesh with molecular biology. Such hybrid names as geochemistry, cosmochemistry, environmental chemistry, agricultural chemistry, food chemistry, and fuel chemistry attest both to chemistry's widespread usefulness and its close ties to other fields of science and technology. The content of this book illuminates chemistry's diversity with discussions of oceanic chemistry, archaeological chemistry, nutritional chemistry, consumer chemistry, forensic chemistry, and human chemistry.

Of special importance is the economic magic of chemistry, one of the themes in the book. The central importance of chemistry's contributions to our economic well-being is apparent in the materials that we use to clothe, house, and transport ourselves, to protect our structures, to produce and package our foodstuffs, and to produce countless other commodities of commerce. Chemistry is central to recovering our mineral resources, finding wider uses for abundant but underutilized mineral resources, developing substitute materials, finding needed sources of energy, and manufacturing a myriad of useful, needed, and wide-ranging products from our petroleum supplies.

Food chemists are studying ways to manipulate the enzymes and chemical reactions which give foods their unique nutritional values, tastes, colors, textures, and storage properties. Another goal is to develop new ways of increasing food production to meet the world's growing problem of starvation among much of its population.

Advances in medicinal chemistry have led to the synthesis of a multitude of compounds that can either cure diseases or alleviate effects and symptoms. The marked strides made in the nation's health through better drugs and better therapeutic techniques, together with much improved diets and sanitation, are reflected in the greatly increased life expectancy that we enjoy today.

I believe that in the future most progress will be made in the chemistry of life processes — in biochemistry, molecular biology, and related areas concerning the study of proteins, enzymes, nucleic acids, and other macromolecules. Chemical and biological investigations at the molecular and cellular levels, aided by enormously efficient computers, will elucidate the origin of life and perhaps lead to the artificial creation of life. Biochemical genetics will give us a great deal of control over the genetic code and, beneficently applied (which will pose a real challenge), should result in a reduction or elimination of genetic defects. Immunochemistry, computer-aided molecular medicine, and chemotherapy should lead to the alleviation, treatment, cure, or prevention of our major ailments, including "mental" illness, and to a slowing of the aging process. Biochemical engineering should make available implantable (microcomputer-assisted) artificial hearts, kidneys, eyes (instruments to permit the blind to "see"), ears (instruments to permit the deaf to "hear"), and other bodily parts and organs.

Because science and technology play such an important role in our lives today and because chemistry can correctly be identified as "The Central Science" a broad public understanding of the widespread contributions of this discipline is essential. The editors of this book have assembled an impressive group of essays that should help the reader, from whatever walk of life, to understand better this central force in our society. Such an understanding can lead to a better life through an increased ability to profit from and to help shape the future course of that force.

Lawrence Berkeley Laboratory
Berkeley, California
January, 1983

Dr. Seaborg, who shared the 1951 Nobel Prize in Chemistry with Edwin M. McMillan, originated the actinide concept for placing the heaviest elements in the periodic system and is the co-discoverer of elements 94 through 102 and 106. Active in governmental and university affairs, he was Chairman of the U.S. Atomic Energy Commission (1961-71), Associate Director of the Lawrence Berkeley Laboratory (1954-61, 1971-present), and Chancellor of the University of California (1958-61). He has been President of the American Association for the Advancement of Science (1972) and the American Chemical Society (1976), has been elected an honorary member of many scientific societies, both domestic and foreign, and is the recipient of numerous honors and awards.

To our wives,
Laurie Kauffman and Nita Szmant

THE CENTRAL SCIENCE

COSMOCHEMISTRY

The Origin of Life in the Universe

Cyril Ponnamperuma

Cyril Ponnamperuma is a native of Sri Lanka (Ceylon). He came to the United States in 1959, and in 1967 he became a naturalized U.S. citizen. He obtained a B.Sc. (Honors) degree in Chemistry at Birkbeck College, University of London in 1959, and his doctorate in chemistry in 1962 under the direction of Professor Melvin Calvin. In 1963 he joined NASA's Exobiology Division and became Chief of the Chemical Evolution Branch. He has been closely involved with NASA in the Viking and Voyager programs. In September, 1971 he joined the University of Maryland as Professor of Chemistry and Director of the Laboratory of Chemical Evolution. The author of over 250 publications related to chemical evolution and the origin of life, he has written and edited a number of books on the subject, including Origins of Life. *He is on the editorial board of the* Journal of Molecular Evolution *and is Editor-in-Chief of the international journal* Origins of Life.

How did life begin? This is a question which has been in the mind of the philosopher and the theologian for ages. But today researchers dare to approach this subject in a strictly scientific and experimental manner. How did this change in attitude take place? Three reasons lead to this. First, the information available from modern astronomy: planets are plentiful in the universe. Each star seen twinkling in the night sky may have around it a planet suitable for life. If that is indeed the case, the number of sites for life in the universe is literally astronomical. Al Cameron has made some calculations in this regard and tells us that 50% of all stars in the universe must have conditions suitable for life. Others who have been less optimistic have given the figure of

5%. Astronomers are agreed, however, that conditions suitable for life appear to be commonplace in the universe.

From astronomy we turn to biochemistry. The nucleic acids (DNA and RNA) and the proteins are at the basis of all life. DNA is a large molecule made up of many nucleotides, but the components are few in number: adenine, thymine, guanine, cytosine, a single sugar, deoxyribose, and a phosphate. In RNA thymine is replaced by uracil and the deoxyribose by ribose. Thus, five bases, two sugars, and a single phosphate account for the variety seen in heredity today. Similarly, if we examine the protein molecule, although it may be very large in molecular weight, it is made up of only 20 amino acids. Between the nucleic acids and proteins, we have 28 components. The alphabet of life is a simple one, only two letters more than that of the English alphabet. Whether we examine an advanced intelligent being, an elephant, or the smallest microbe, we cannot get away from the fact that the nucleic acids and the proteins are at the basis of all life. Biochemically speaking, it is the interaction of these two kinds of molecules which give us the phenomenon of life. Therefore the inescapable conclusion is that all life must have had a common chemical origin.

From biochemistry we turn to the third basis for the scientific study of this problem, namely the concept of Darwinian evolution, which is the cornerstone of modern biology. If we accept Darwinian evolution, we must postulate an earlier form which we call "chemical evolution." This idea was expounded in 1871 by the British physicist Tyndall in his book *Fragments of Science for Unscientific People:*

> Darwin placed at the root of life a primordial germ from which he conceived that the amazing richness and variety of life now upon the Earth's surface might be deduced. If this hypothesis were true, it would not be final. The human imagination would infallibly look behind the germ and, however hopeless the attempt, inquire into the history of its genesis. A desire immediately arises to connect the present life of our planet with the past. We wish to know something of our remotest ancestry. Does life belong to what we call matter, or is it an independent principle inserted into matter at some suitable epoch when the phys-

ical conditions became such as to permit the development of life?

Darwinian evolution logically leads all the way back to the very birth of the universe. We can trace a continuum from the atoms that were formed during the birth of a star to the appearance of man.

This idea of the continuity of life is an ancient one that has been expressed in many ways and in many languages. The philosophers of old gave us the concept of spontaneous generation. The *Rig Veda* and *Atharva Veda* of ancient India describe the oceans as the cradle of all life. Aristotle in his *Metaphysics* suggests that "fireflies arose from morning dew." The world's literature is full of references to spontaneous generation. Vergil describes "a swarm of bees arising from the carcass of a calf." We have even a reference to "the crocodile of Egypt born of the mud by the action of the sun," in Shakespeare's *Antony and Cleopatra*. This concept was accepted by most prominent medieval thinkers; Newton, Harvey, and van Helmont also subscribed to the theory of spontaneous generation.

It is amusing that even in the seventeenth century we note an irresistible urge for the making of mice in the writings of the Belgian physician, van Helmont: "If a dirty undergarment is squeezed into the mouth of a vessel containing wheat, within a few days, 21 is a critical period, the ferment drained from the garment and transformed by the smell of the grain encrusts the wheat itself with its own skin and turns it into mice. And what is more remarkable, the mice from corn and undergarments are neither weanlings nor sucklings, nor premature, but they jump out fully formed."

But such ideas could not long withstand the advancing rigors of scientific thought. Louis Pasteur dealt the death blow to the idea of spontaneous generation. In 1861, describing his experiments before the French Academy, Pasteur made this dogmatic statement: "Never will the doctrine of spontaneous generation arise from this mortal blow." We hold out Louis Pasteur's experiments to our beginning students in the natural sciences as a triumph of reason over mysticism. But today we are returning to spontaneous generation. However, the spontaneous generation we propose is not mice from old linen, nor frogs from a primordial ooze, but rather an orderly sequence of events starting with the formation of the elements at the beginning of the universe. We move from atoms to small molecules, from these, to larger molecules, and finally to the replicating system which is now recognized as the basis of all life. Every erroneous idea may have a germ of truth in it, and today we present the hypothesis of chemical evolution as a refinement of the ancient doctrine of spontaneous generation.

As we approach the modern scientific thinking on the subject of the origin of life, we find that one of the first to express himself was Charles Darwin. In the now-celebrated letter to his friend Hooker in 1861, he wrote, "If we could conceive, in some warm little pond with all sorts of ammonia and phosphoric salts, light, heat, electricity, et cetera, present, that a protein(e) compound was chemically formed ready to undergo still more complex changes." This is at the root of all our thinking on the origin of life.

The experimentalist in chemical evolution attempts to recreate Darwin's warm little pond in order to examine whether reactions that we postulate for the beginning of life can be reproduced under simulated conditions in the laboratory.

The raw material available to us is the material of the sun, a close approximation of the average composition of the solar

TABLE I. Composition of the Sun

Element	Abundance (%)
Hydrogen	87.0
Helium	12.9
Oxygen	0.025
Nitrogen	0.02
Carbon	0.01
Magnesium	0.003
Silicon	0.002
Iron	0.001
Sulfur	0.001
Others	0.038

system. Hydrogen, oxygen, nitrogen, and carbon — the most abundant elements if we omit the inert gas helium — are the most important elements for life. These elements make up 99½% of the biosphere (Table I). In the presence of hydrogen, the carbon, the nitrogen, and the oxygen would be converted (or "reduced") to methane, ammonia, and water, respectively. It is reasonable to expect the presence of intensely reduced materials during the very early stages of the development of planets from the primordial nebula. These are the raw materials with which we have to work. We owe a great debt to Alexander Ivanovich Oparin, who in 1924 gave us the concept of chemical evolution in scientifically defensible terms by suggesting a hydrogen-laden or reducing atmosphere as a starting point. He did this four years before Russell discovered the abundance of hydrogen in the universe. Oparin's conclusion was based primarily on biochemical requirements for the reduction of carbon to produce the organic molecules necessary for life.

Our available energy sources are listed in Table II. We have first the light from the sun. This was indeed the most abundant source of energy available. In the early stages of prebiological chemistry, short wavelength light was perhaps more important than the rest of the solar spectrum in terms of electrical discharges. Although the total amount of short wavelength light close to the earth was small, these short wavelengths may have been rather effective in the introduction of organic compounds into the primordial ocean. Next are the cosmic rays and radioactivity in the crust of the earth or dissolved in the oceans, per-

TABLE II. Energy Available for Synthesis of Organic Compounds

Source	**Energy** (cal cm^{-2} yr^{-1})
Ultraviolet light (2500 Å)	570.
Electric discharges	4.
Radioactivity	0.8
Heat from volcanoes	0.13

haps in the form of the radioactive isotope potassium 40. The dissolved radioactivity in the oceans could have been an effective tool for the synthesis of organic compounds. The heat from volcanoes and the shock waves generated during a meteorite impact may also have played a role. We may say that not only lightning but also thunder was responsible for the synthesis of these organic compounds.

The chemical evolutionist attempts to trace the path from a primitive atmosphere to nucleic acids and proteins. Though we have presented the primitive atmosphere in its most intensely reducing form, one can also consider carbon monoxide and carbon dioxide as sources of carbon compounds. The nitrogen of ammonia could be replaced by nitrogen gas, and the water can be not only a source of oxygen but also of hydrogen.

The experimental work can be divided into two parts: first, to see if these processes give rise to the monomers of biological significance, and, then, to see if the same kind of energy sources give rise to the polymers necessary for life. We label this the synthetic approach to the problem of the origin of life. We attempt to make the molecules necessary for life and see if we can put them together to generate the replicating system which is the basis of all life.

The kind of apparatus that is used in the laboratory is illustrated in Figure 1. The upper flask represents the atmosphere; the lower flask represents the ocean. The side arm is kept hot, and the condenser between the upper and lower flasks is kept cold. The circulation that is established portrays the interaction between the atmosphere and the hydrosphere. In the apparatus shown in Figure 1, the source of energy is electrical discharges, but the energy source could be ionizing radiation, heat, shock waves, or any other source of energy.

At the end of a 24-hour experiment, 95% of the methane present at the start of the experiment is converted into organic compounds, which appear as a dark brown material that may be analyzed for chemicals of biological significance.

We should mention here the ideas of Haldane. In 1928, Haldane, independently of Oparin, suggested that the ultraviolet light acted on an early atmosphere and produced a hot dilute soup. The oceans to him were nothing but a soup of organic

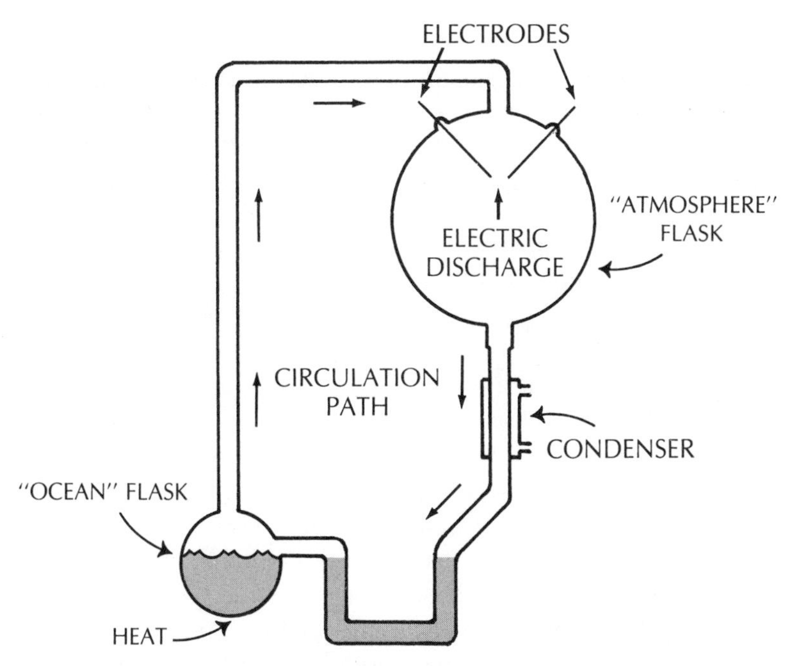

FIGURE 1: Apparatus for Synthesizing the "Primordial Soup"

chemicals. Some call it a hot soup, others think of it as a gazpacho. Whatever the case, it was a solution of organic compounds. We make the "primordial soup" in the laboratory in many ways for purposes of analysis.

In place of electrical discharges one can also use ultraviolet light from a very powerful ultraviolet source. If one uses a 15-atmosphere argon source employed by NASA to irradiate spacecraft materials, then, because such a lamp emits 10^{19} photons/cm^2/sec while the sun's intensity in this region of 1,000 to 2,000 Å is only 10^{15} photons/cm^2/sec, one can simulate in a short time what may have happened over an extended period on the primitive earth. Also, ionizing radiation can be used as a source of energy. In such experiments, one discovers the formation of adenine. Adenine is one of the single most important biochemicals. The fact that a single, random process produces such a compound has a great impact on the thinking concerning the effect of primordial energy sources on a mixture of very simple chemicals.

Thanks to the painstaking work of several organic chemists, the mechanisms required for the production of adenine are now understood. It turns out that the structure of adenine consists of five molecules of hydrogen cyanide $(HCN)_5$. $(HCN)_5$ has been detected by radio astronomers in the interstellar space. The same is true of the simple molecule formaldehyde (H_2CO). This substance is also produced when a mixture of methane, water, and the other simple carbon compounds are exposed to shock waves. Hydrogen cyanide and formaldehyde are the key molecules necessary for the generation of the organic compounds which are the building blocks of the nucleic acids and the proteins.

Thus, experiments in which primitive atmospheres — methane, ammonia, water — are exposed to various kinds of energy give us most of the molecules necessary for the assembly of nucleic acids and proteins. There is no doubt that one can make these molecules in the absence of life. One can then proceed to the next stage of the problem: making polymers from monomers or joining simple molecular structural units into large assemblies.

In putting two amino acids together, we must remove a molecule of water, and a di-peptide is thus formed by a dehydration condensation. Such a dehydration could have taken place on the primitive earth by the evaporation of the primordial soup along the ocean shoreline. In his celebrated lecture to the British Physical Society entitled "The Physical Basis of Life," Bernal suggested that organic components from the ocean were brought from the ocean to the shoreline, absorbed in clay, and there the condensation took place. The Bernallian idea has been tested in the laboratory with encouraging results.

Yet another possibility is that the reaction could take place in water. Since the earth is a very wet planet, if one can demonstrate that this reaction does take place in water, a greater opportunity would exist to examine the formation of polymers under prebiotic conditions. Removing water in water is a difficult task. To put it very naively, it is like trying to keep dry while swimming. But we know that processes in living systems have evolved ways and means of doing so. There are enzymes or biochemical catalysts that can act as energy sources to accomplish this difficult task. One must look for the predecessors of these modern enzymes, the primitive catalysts and the phosphates.

In this context it is noteworthy that the substance phosphine (PH_3) is claimed to be present in the Jovian atmosphere. If phosphine were present in the primitive atmosphere, the chances of it being transformed to phosphate are likely. This reaction may have lasted long enough in the early oceans so that the phosphate gave rise to linear polyphosphates, some of the most effective condensing agents. The polyphosphates could also have been generated during the course of heating. It is generally believed that the earth must have reached at least 1000° C. during the stages of accretion, i.e. the accumulation of matter during the formation of the planet. If the oceans were formed by the outgassing of the crust, then the polyphosphates would have been leached into the oceans and could have functioned as effective condensing agents.

An additional source of polyphosphates is suggested when one considers planetary formation. During late accretion, even after the oceans were already formed, there was much scavenging in the solar system, and some of the debris from the falling primordial dust, such as Scribezite, could have produced linear polyphosphates.

Let us approach the problem of the origin of life from a chronological point of view. Figure 2 is the geologic clock. The processes that took place on earth are represented on a 12-hour diagram modified after the work of Schopf of UCLA. At one time, no one thought that life existed before the base of the Cambrian period, 600×10^6 years ago. No skeletons had been found, and it was argued that there was no life. However, thanks to the painstaking and careful work of micropaleontologists, notably Barghoon and his co-workers, we now can trace life all the way back to one billion years ago in the Bittersprings formation in Australia and to two billion years ago in the Swaziland sequence. A paper by Barghoon and Knoll (*Science*, 1977) examined the 3.4-billion-year-old Swartkoppie formation of South Africa. They claim to have found microstructures which appear

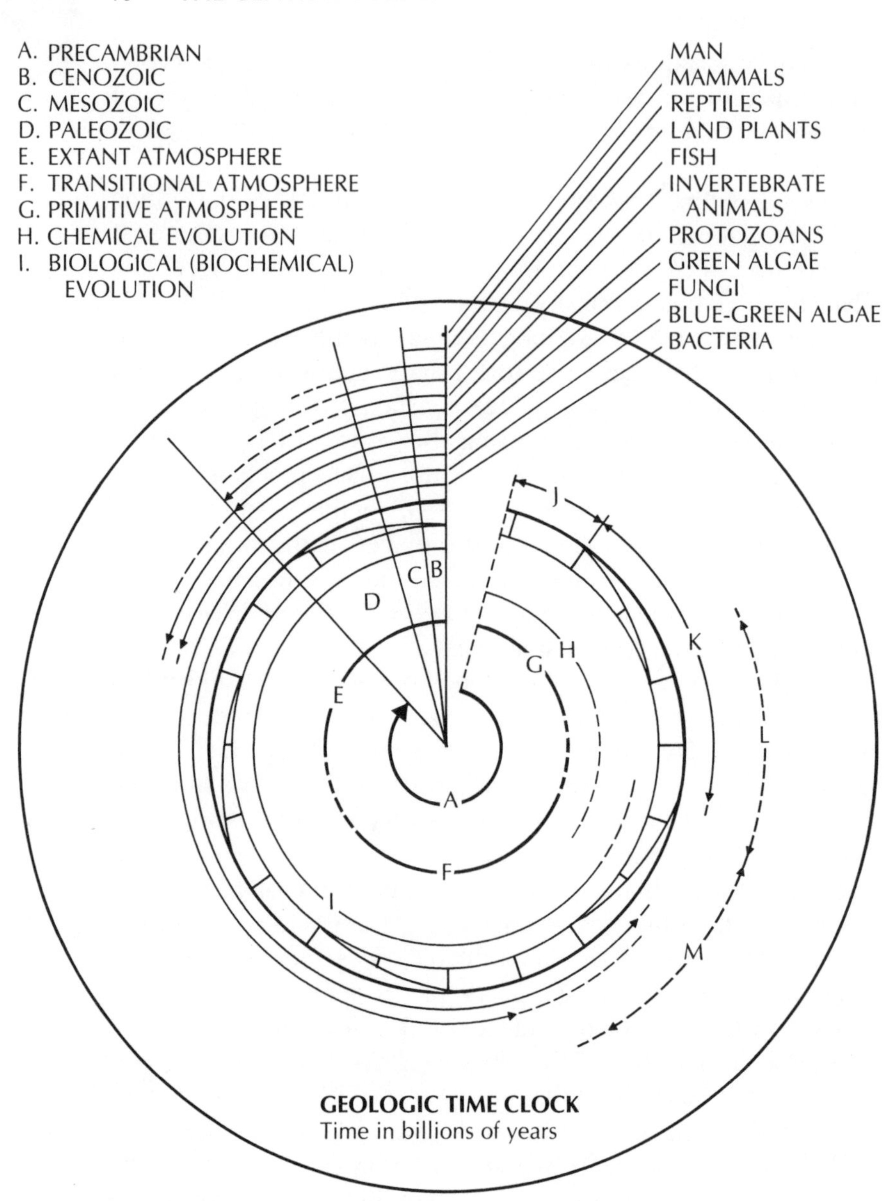

J. ORIGIN OF EARTH
K. ORIGIN OF CRUST AND CORE
L. BEGINNINGS OF LIFE PROCESSES
M. BEGINNINGS OF PHOTOSYNTHESIS

FIGURE 2: Geologic Clock, Modified after Schopf (1967)

to be algae-like. If this is indeed true, then we have dated the oldest life on earth at 3.4-billion years ago. More recently, stromatolites have been found in rocks 3.5 billion years old.

Lunar samples have been analyzed in the laboratory with disappointing results. In the 200 ppm of carbon found in lunar samples there were traces of hydrocarbons, but no amino acids or nucleic acid bases or sugars were detected. About 2,000 meteorites are present in the collections around the world. Of these, 36 fall into the category known as carbonaceous chondrites; they contain organic materials from half a percent to about five percent of organic material. A celebrated carbonaceous chondrite is the Orgueil meteorite which fell on May 12, 1864. This was analyzed by Berzelius, who reported six percent of organic matter. Today's analysis reveals five and a half percent, a figure not too far from the original report.

A meteorite which provided the first convincing proof of extraterrestrial amino acids is the Murchison that fell on Sept. 28, 1969, in Australia. Analysis of the meteorite showed that it contained a large number of amino acids, protein, and nonprotein material. In structures that have asymmetric centers, both the D- and L-forms were found. There are two aspects to work with here, the presence of the D- and L-forms and the nonprotein amino acid content. A large body of information has been accumulated, and it indicates that many of the molecules necessary for life processes are indeed present in the Murchison.

Recently, U.S. and Japanese polar scientists have returned from the Antarctic with over 3000 meteorites. Meteorites appear to accumulate on the blue ice of the Antarctic, where the conveyor belt action seems to bring them to the terminal moraines. Three of these meteorites appear to be carbonaceous chondrites, and work on these materials is continuing. We may have a further opportunity to confirm some of our ideas about carbonaceous chondrites by using these recent finds.

A great deal of information has accumulated from the Viking mission to Mars. It is clear from the organic analysis data, whether examining the Plains of Chryse or the Plains of Utopia, that the organic component is extremely small, less than 10 parts per billion. This is less carbon than that found on the moon. In the absence of organic matter, the chances of life appear to be remote. Although the results of the biological experiments were apparently positive, they may be explained by surface chemistry. The peroxides on the Martian surface may have mimicked the microbes, and we must conclude very reluctantly that the experiments performed by the Viking lander show no evidence of the presence of life on Mars. Perhaps one should revise the question and ask, "Was there life on Mars?" This approach may be more useful in searching for answers to the question of the origin and evolution of life from planetary exploration.

In our solar system, the only place where life exists is the earth. To answer the question, "Are we alone in the universe?" we may have to go beyond our solar system. If life began on earth, evolved to the present form, and if intelligence developed from it, then perhaps the same thing may have happened or may be happening elsewhere in the universe.

If we ask our radioastronomers how many civilizations there are in the universe, they will give us the answer $N = R_* f_p f_e f_l f_i f_c L$, where R_* = the rate of star formation; f_p = the number of stars with planets; f_e = the number of those planets which have an ecology suitable for life; f_l = the fraction of those planets on which life has evolved; f_i = number of cases in which that life has evolved to the point of intelligent life; f_c = the number of times that intelligent life has become communicable; and L = the longevity of a civilization.

So far, it is estimated that in our galaxy there may be a million civilizations.

Six searches for extraterrestrial life are currently being conducted via radiotelescope. At the University of Maryland, Professor B. Zuckerman is using the Greenbank Observatory. There is one in Ohio, one in Arecibo, Puerto Rico, and two in the Soviet Union. Before long a distant "hello" from a galactic neighbor may tell us that we are not alone in the universe.

SUGGESTED READINGS

BOOKS

Bernal, D. *The Physical Basis of Life*. London: Routledge and Paul, 1951.

Field, G., G. Verschuir, and C. Ponnamperuma. *Cosmic Evolution*. Boston: Houghton Mifflin, 1978.

Oparin, A. I. *The Origin of Life*. New York: Dover Publications, 1938.

Ponnamperuma, C. *Exobiology, A Series of Collected Papers*. London: North Holland, 1972.

____________. *The Origins of Life*. New York: E. P. Dutton, 1972.

____________ and A. G. W. Cameron (eds.). *Interstellar Communication: Scientific Perspectives*. Boston: Houghton Mifflin, 1978.

ARTICLES

Gammon, R. H. "Chemistry of Interstellar Space." *Chemical and Engineering News*, 56, No. 40 (October 2, 1978): 21-33.

Langone, John. "Cyril Ponnamperuma: Meteorites and the Stuff of Life." *Discover* 4, No.11 (November 1983): 50,51,54,57,60.

Milligan, D. O. (ed.) "Cosmochemistry." *Proceedings of the Robert A. Welch Foundation Conferences on Chemical Research,* Houston, Texas, 1978.

Ponnamperuma, C. "Life Beyond the Earth." *Astronautics and Aeronautics*, 14, No. 11 (November 1976): 50-55.

CHEMISTRY AND THE OCEANS

Understanding Our "Water Planet"

Philip N. Froelich, Jr.

Philip N. Froelich is Associate Professor of Oceanography at Florida State University, Tallahassee, Florida. A 1979 graduate of the Graduate School of Oceanography at the University of Rhode Island, he is the author of several dozen research papers on the geochemistry of nutrients and trace elements in seawater, in estuaries, and in marine sediments and their interstitial waters. During the last 13 years he has spent more than 10 months at sea on oceanographic research voyages, more than two and a half months as chief scientist of major expeditions in the Atlantic and Pacific Oceans.

If there were astronomers on Venus or Mars, they would undoubtedly call Earth the "water planet," for, seen from space, our globe sparkles like a drop of glistening dew. Only an occasional drab brown patch of continent mars the vast expanse of the ocean. Even though water covers almost 70% of the surface of this planet, until fairly recently we knew less about the ocean bottom than we knew about the moon's surface. During the last three decades, however, a major effort has been made by scientists in many disciplines to increase our understanding of the ocean. The result has been a virtual explosion of knowledge, dramatically emphasizing the central role that the ocean plays in such diverse areas as weather and climate, the composition of the atmosphere, shaping of the continents, and the formation of economically important minerals both on the continents and on the sea-floor. In fact, interactions between the sea, land, and air are so important to our understanding of the physicochemical, biological, and geological forces around us that studying these interrelationships has become a major part of present research in the earth sciences.

Chemistry provides many vital tools in these research programs. Marine chemists are uniquely suited to investigate the interchange of elements (the basic building blocks of all matter) between the oceans and other reservoirs of the earth's surface (the atmosphere, the biosphere, the continents, and the seafloor). Chemists study the reactions which occur during erosion of the continents and the transport of both dissolved and particulate elements to the sea by rivers. They study the fate of these elements in the oceans, how they are carried and mixed by ocean currents and eventually removed from seawater and deposited in ocean-floor sediments, and how these marine sediments are finally compressed, molded, folded, and raised into continents to be eroded again and carried down to the sea several hundred million years later. The salts dissolved in seawater are the direct result of this cycle of continental erosion, riverine transport, marine sedimentation, and continent formation.

Many oceanographic researchers use chemical tools even though they may not consider themselves chemists. Geological oceanographers use chemical techniques to characterize marine sediments and measure the rates at which sediments accumulate. Physical oceanographers employ a wide variety of chemicals to trace the movement and mixing of water masses within the oceans. Marine biologists and ecologists use chemical methods to measure the growth rate of plants in the sea and to study the uptake of nutrients by marine plants and the transfer of food and energy between living systems and their environments. They can measure how fast a coral grows or determine the effects of pesticides and other pollutants on marine life with chemical means. In short, a list of all the chemical analyses being done on the oceans would be almost endless.

Three broad areas of ocean chemistry demonstrate the interdisciplinary nature of oceanography and the importance that marine chemists must place on understanding the physical, geological, and biological processes affecting the ocean's chemistry. They are: (1) oceanic mixing and its effect on atmospheric carbon dioxide; (2) life in the sea and its effect on ocean chemistry; and (3) chemical interactions between seawater and undersea lavas during sea-floor spreading.

THE CARBON DIOXIDE (CO_2) PROBLEM

CO_2 is a key ingredient in the earth's heat balance. Short wavelength sunlight passes through the atmosphere and warms the earth. Long wavelength heat waves given off by the heated earth are absorbed by atmospheric CO_2 and are prevented from escaping into space, thereby holding in some of the sun's heat, a phenomenon called the "greenhouse effect." The average temperature of the earth is dependent upon the efficiency of this CO_2 heat trap: the higher the CO_2 concentration, the higher the earth's temperature.

The atmospheric CO_2 concentration has increased almost 15% since about 1880. This increase has been caused by two man-made effects: the burning of fossil fuels (coal, oil, and gas) for energy and the clearing of forests for firewood and agricultural use. About one-half of the CO_2 released by these two processes has been taken up by the oceans during the last one and one-half centuries. Otherwise, atmospheric CO_2 would have increased by twice as much — more than 25%. Within the next century it is expected that the world's industrial society will burn enough fossil fuel to double or quadruple the atmospheric CO_2 content, depending upon how much and how fast fossil fuel is burned and how fast the oceans can continue to absorb the excess.

Doubling of the atmospheric CO_2 may cause a rise in the earth's average temperature of about 2° to 3° C. This does not seem to be much, but a number of lines of evidence suggest that it could cause climatic changes with serious environmental and economic consequences for future generations. We can anticipate altered climates in many areas, shifts in average rainfall patterns and perhaps a rise in world-wide sea level of about three meters (9.8 feet) over hundreds of years. This rise in sea level, caused by melting glacial ice in Greenland and Antarctica, would be enough to flood low-lying areas of the major coastal cities of the world. At worst, the shifts in rainfall patterns may disrupt our planet's already overburdened agricultural system because our richest lands are located in temperate regions that today receive ample rainfall. The immediate threat of the CO_2

problem is not a cause for alarm, but it should be a matter of concern to the world's societies in making long-range plans for potential energy sources for future generations.

OCEANIC MIXING

The ocean has an almost limitless capacity to absorb excess CO_2 because of its enormous volume. Unfortunately, however, the rate at which excess CO_2 can be absorbed is limited by the rate at which the surface waters in contact with the atmosphere can mix downwards and be replaced at the surface by new seawater ready to take up its share of the excess CO_2. This depends on how fast the ocean mixes.

In terms of mixing, it is useful to think of the ocean as two separate bodies of water: one, a thin, shallow warm-water ocean, and the other, a large, deep, denser, cold-water ocean. The shallow, warm-water ocean is in constant contact with the atmosphere and takes up excess CO_2 rapidly. However, since it makes up only a small percentage of the total volume of the oceans, it has a limited capacity to absorb excess CO_2. The deep, cold-water ocean, which comprises over 90% of the ocean's volume, cannot effectively take up excess CO_2 because it is prevented from mixing with the overlying shallow ocean by a thick thermal boundary, the thermocline, between them. The only pathway for rapid gas exchange between the atmosphere and the deep sea is in the polar regions where this thermal barrier is broken down, permitting mixing between the shallow and deep-sea layers.

Marine chemists have estimated mixing times for these two oceanic reservoirs by taking advantage of a naturally-occurring and man-made radioactive "clock": carbon-14 (radiocarbon). Radiocarbon is chemically identical to its more abundant sister element, carbon-12, which is the ordinary carbon found in wood, coal, and living organisms. Unlike stable carbon-12, carbon-14 decays radioactively in a predictable manner, providing us with a sort of built-in timer in the CO_2 system.

Trace amounts of carbon-14 exist naturally in the atmosphere. It decays to form ordinary nitrogen-14 with a half-life (the time required for one-half the atoms of a given amount of a radioactive element to disintegrate) of about 6000 years. In other words, if we isolated one hundred carbon-14 atoms from the atmosphere, half of them would disappear every 6000 years. After 6000 years we would have 50; after 12,000 years, 25; and so forth. The radiocarbon content of the atmosphere (as radiocarbon dioxide) was constant up to about the year 1800. After 1800 (the beginning of the Industrial Revolution), fossil-fuel CO_2 was added to the atmosphere. Fossil-fuel carbon is millions of years old and thus contains no radiocarbon, which decays completely within about seven half-lives (45,000 years) after the fossil deposit was formed. Thus we have been diluting atmospheric radiocarbon dioxide with "dead" fossil-fuel carbon dioxide. (This decrease in radiocarbon is called the Suess effect.) For the period prior to 1800, we can trace the rate at which radiocarbon dioxide has mixed into the deep-sea by measuring the decay of carbon-14 in the deep–ocean waters. The amount of missing carbon-14 tells us how old the water is or how long it has been since it was last in contact with the atmosphere.

Such measurements indicate that the average water sample in the deep sea comes in contact with the atmosphere only once every thousand years or so. This slow rate of mixing suggests that the deep-ocean reservoir can only take up excess CO_2 very slowly, on time scales of thousands of years. Detailed calculations indicate that during the next century, when most of the excess CO_2 production will occur, only about 15% of the excess CO_2 will be transported into the deep sea through the thermocline in polar regions.

We can use carbon-14 in a different way to estimate mixing and CO_2 transport in the shallow ocean. During the nuclear bomb tests of the 1950s and 1960s, a large quantity of "bomb" radiocarbon was added to the atmosphere. This radiocarbon has not yet had time to decay perceptibly, but the increased concentration is readily measurable in the atmosphere and in those parts of the shallow ocean that have exchanged CO_2 with the atmosphere during the last 20 years. Measurements of this bomb carbon-14 in surface waters suggest that some radiocarbon has penetrated through the surface layers into the upper thermocline separating the shallow and deep oceans. These estimates, plus evidence from other radiotracers, suggest that about 35% of the excess CO_2 produced over the next century will be absorbed into the shallow ocean and the upper layers of the thermocline.

Thus, if our mixing estimates are correct, only about one-half of the total fossil-fuel CO_2 production expected in the next one hundred years will be transferred to the oceans. The remainder is likely to accumulate in the atmosphere, a potential threat to our climate. Oceanic mixing rates, however, are based in part on imperfect mathematical models of the ocean plus a scattering of hard data. Thus our present estimates of future excess CO_2 levels are tentative, and much effort is being devoted to developing better models and collecting more data. Recently, it has been suggested that forest ecosystems may be important in taking up and releasing excess CO_2 during time scales of decades. This suggestion has strengthened the realization that scientists from many fields (oceanography, atmospheric chemistry, terrestrial ecology, etc.) must work together to understand the future impact of the CO_2 problem.

LIFE IN THE SEA, A SELF-FERTILIZING GARDEN

The oceanic mixing process and the rate at which deep-ocean water is mixed back to the surface are also of critical importance to marine life. The oceans are illuminated to a depth of only one hundred meters or so (about 300 feet) by the sun's light. Below this, several thousand meters of the oceans' depths are eternally dark and incapable of supporting any form of plant life. For this reason, almost all plant and animal life in the sea is concentrated near the surface: plants because they need sunlight and animals because they feed on the plant life there. This life is composed primarily of microscopic organisms that live suspended in the shallow ocean, drifting about wherever oceanic currents may carry them. As they multiply and grow, they take essential nutrients such as nitrogen and phosphorus from the water. This continues until the concentrations of these two "fertilizer elements" in the surface water are almost zero, a condition which limits further growth.

When organisms die or are eaten, their biogenic remains sink slowly from the surface into the deep ocean where they decompose (Figure 1). The decomposition process releases nutrients directly into the deep water. Oceanic mixing slowly transports nutrient-rich deep water back into the surface waters where it again fertilizes growth and production of new marine life. This new life then dies, sinks, and decomposes, adding its nutrients to the deep sea so that the nutrient cycle is continuous. In this sense, the ocean is a perpetual self-fertilizing garden (Figure 1). If the ocean were ever to cease mixing, it would become a

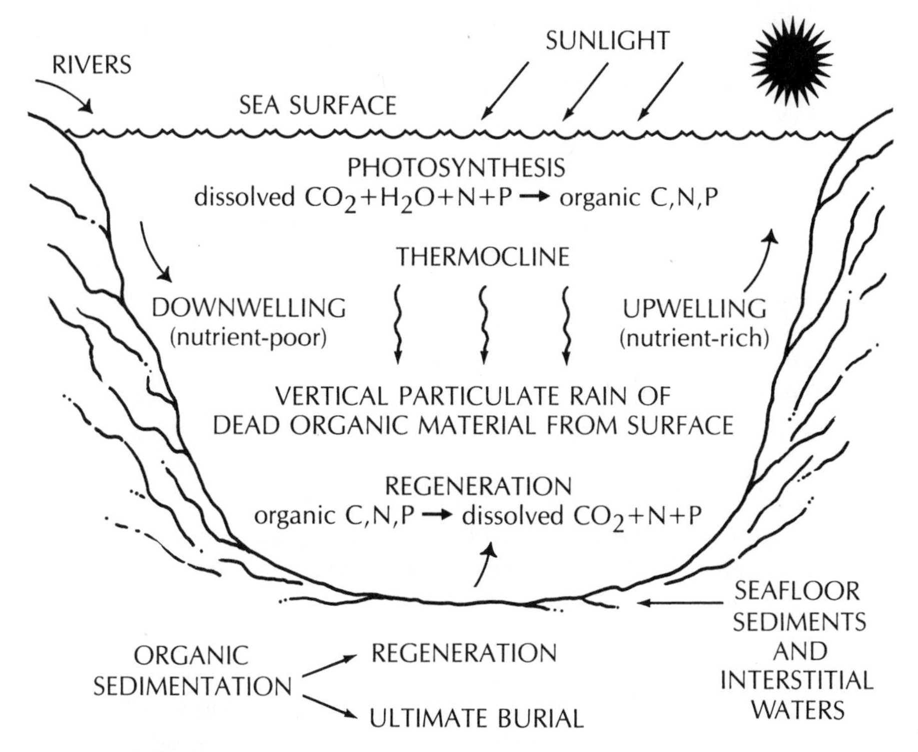

FIGURE 1: The self-fertilizing action of the oceans. Nutrients (N and P) are utilized by phytoplankton in the surface ocean during photosynthesis to produce living organic material which either dies or is eaten and sinks out of the euphotic zone (the upper 100 meters — 300 feet — of sunlit water). In the deep sea, organic matter decomposes and releases nutrients back to the deep sea. Some regeneration occurs in the upper sediments. The nutrient-rich deep water is then brought back to the surface ocean by upward mixing in upwelling zones, which are areas of high productivity. This upwelled water is then stripped of nutrients by biological activity and eventually downwells in areas of deep- and bottom-water formation, carrying nutrient-depleted water into the deep sea.

desert, for the life-giving nutrients would become trapped in the deep sea.

Not all the dead organic matter falling from the surface ocean is regenerated in the water column. A small portion escapes destruction and is buried in the sediments, where the decomposition process continues. But here, the nutrients are released into the interstitial water between the sediment particles rather than into the overlying seawater; the nutrients become trapped and build up to much higher concentrations. These nutrients leak back into the overlying seawater so slowly that it takes about a thousand years for a nutrient molecule released six feet deep in the sediment to migrate up to the sediment surface and to mix upward with deep-ocean water.

By studying the concentrations of nutrients in the deep ocean and in interstitial waters, marine chemists can estimate how much organic debris is being regenerated and how important the slow release of nutrients from sediments is to the plant productivity of the ocean. Such investigations help us to understand the interrelationships between the biological and chemical mechanisms at work in the sea and someday may allow us to take better advantage of the food resources in the ocean.

SEAWATER AND SEA-FLOOR SPREADING

Marine chemists have recently become interested in an area that was previously the geologists' domain: sea-floor spreading. During the last twenty years, marine geologists and geophysicists have discovered that the deep-sea floor (the oceanic crust) is being constantly created along long, linear undersea mountain ridges that stretch down the middles of the world's four oceans. The oceanic crust on both sides of these ridges is moving in opposite directions away from the crest at speeds of several centimeters per year (one centimeter = ⅜ inch). Hot molten magma from within the earth pushes upwards under these mid-ocean ridge crests and occasionally breaks through. These undersea lava flows cool and solidify to form new crustal rock that replaces that which has moved away. The result is analogous to two back-to-back conveyor belts carrying the seafloor away from the mid-ocean ridges on either side, while new sea floor is being created at the middle.

Marine chemists are interested in sea-floor spreading because seawater reacts with the hot rocks near the ridge crest (Figure 2). During crustal formation, fissures or cracks form in the cooling lava that permit seawater to percolate down into the hot, newly-formed crust. As the water is heated under pressure, it reacts with the rock, and the chemical composition of the seawater is altered. Many metals are leached from the rock so that the seawater becomes a fluid (hydrothermal solution) barely resembling the original seawater. These hot, metal-laden solutions then escape from the rock through vents and fissures and mix into the overlying seawater. The metals, including high concentrations of iron and manganese, precipitate out rapidly and are deposited near vents in unique muds called metalliferous sediments. These sediments resemble continental metal-rich ore deposits. Many geochemists believe that this same sea-floor hydrothermal process was responsible millions of years ago for the formation of the important ore deposits that are found today

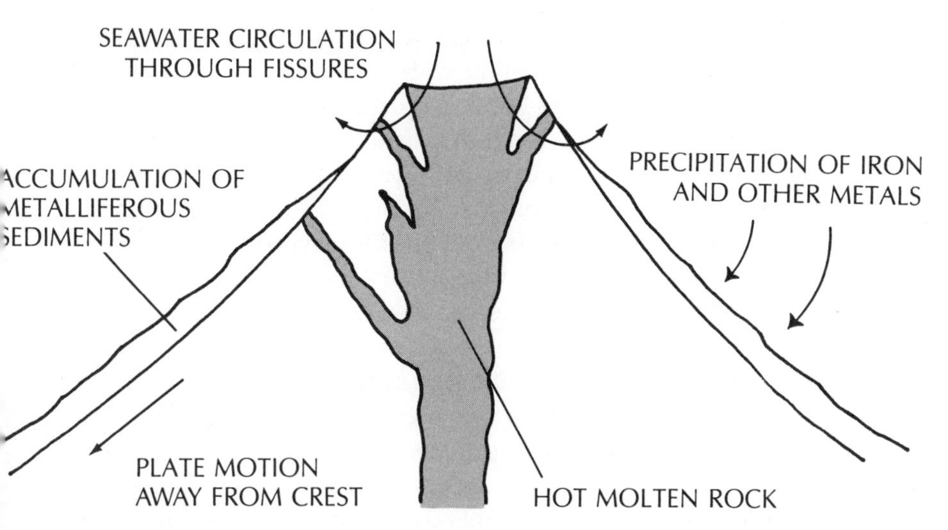

FIGURE 2: Formation of metalliferous sediments near actively spreading mid-ocean ridge crests occurs by circulation of seawater down through fissures in the hot rocks. The heated seawater reacts with the rocks to form a hydrothermal solution that precipitates when the seawater vents into the overlying seawater. The precipitated iron and other metals originating from the hydrothermal solutions and from seawater (scavenged by the iron precipitate) accumulate in nearby metalliferous sediments.

on the continents in the mines that have been worked by man for their valuable metal contents.

In addition, the concentrations of several of the major dissolved salts in the original seawater are changed in the hydrothermal solution because of reactions between seawater and the hot rocks. Since the entire ocean volume passes through the global mid-ocean ridge crest system about once every 10 million years, the contribution of salts to the sea by hydrothermal processes is as important as the role of rivers.

OASIS IN THE ABYSSAL DESERT

During a recent study of one of these undersea hydrothermal vent systems, a group of chemists made a rather unexpected biological discovery. Using "Alvin," a special submarine equipped for scientific work in the depths of the sea, they were the first to visit personally a deep-sea hydrothermal vent in order to observe the hot, altered seawater flowing out and to take samples of hydrothermal solutions (Figure 3). What they did not expect was to find that the rocky terrain within several tens of feet of each vent was teeming with large clams, crabs, worms, and a number of strange-looking creatures that had never been seen before (Figure 4). The immediate question everyone asked was: Why is there so much life around these vents when the surrounding sea-floor is practically barren? The answer was provided by a chemical understanding of the hydrothermal system. The solutions exiting from vents are rich not only in metals but also in hydrogen sulfide, the gas that causes the offensive odor of rotten eggs. A special type of bacteria is able to live by consuming this sulfide. It proliferates in the immediate vicinity of the hydrothermal waters. The bacteria are filtered from the water and eaten by larger organisms, which are then in turn themselves eaten by still larger animals, and so on. The sulfide being released from these vents thus forms the basis for a food chain that supports the entire population of each vent. To our knowledge this is the only marine ecosystem on the face of the earth which derives its ultimate energy source from heat within the earth rather than from the sun via photosynthesis in green plants. Periodically, these hydrothermal vents cease flowing for reasons that are not understood. A number of inactive vents were observed, surrounded only by dead clams and crab carapaces, mute evidence of a previous activity.

These are but a few of the many areas in which chemistry is helping us to understand the oceans. The continuing study of the oceans through chemistry is increasingly dependent upon new chemical research tools and techniques. Some of the most exciting work being done today is being carried out with techniques and instruments that were designed and built by students. The future of oceanography, indeed of any science, is dependent on the creative effort and unshackled enthusiasm of students. Without them, we will understand no more about the oceans tomorrow than we do today, and there is much to be learned.

FIGURE 3: "Black Smoker" vent observed on the crest of the East Pacific Rise at 21° N, just south of the Gulf of California. This site has become known as the "twenty-one degree north" smoker. The vents here are much hotter than those on the Galápagos Rift and discharge a metal-sulfide-laden hydrothermal solution that is black in color. Photograph taken by Dudley B. Foster from the submersible "Alvin" of the Woods Hole Oceanographic Institution. Part of the "Alvin's" peripheral equipment can be seen in the foreground. Courtesy of the National Geographic Society.

FIGURE 4: Teeming life within feet of a hydrothermal vent on the Galápagos Rift. Photograph taken by Kathleen Crane from the "Alvin" of the Woods Hole Oceanographic Institution. Courtesy of the National Geographic Society.

SUGGESTED READINGS

BOOKS

Holland, Heinrich D. *The Chemistry of the Atmosphere and Ocean*. New York: Wiley-Interscience, 1978.

van Andel, Tjeerd. *Tales of an Old Ocean*. New York: W.W. Norton and Co., 1977.

ARTICLES

Ballard, R.D., and J.F. Grassle. "Incredible World of Deep-Sea Rifts." *National Geographic* 156, No. 5 (November 1979).

Corliss, John B., and Robert D. Ballard. "Oases of Life in the Cold Abyss." *National Geographic* 152, No. 4 (October 1977).

Edmond, John M. "GEOSECS Is Like the Yankees: Everybody Hates It and It Always Wins." *Oceanus* 23, No. 1 (Spring 1980): 33-39.

"Galápagos, '79: Initial Findings of a Deep-Sea Biological Quest." By the participants of the original team which discovered hydrothermal vents. *Oceanus* 22, No. 2 (Summer 1979): 2-10.

Koski, Randolph, *et al.* "Metal Sulfide Deposits on the Juan de Fuca Ridge." *Oceanus* 25, No. 3 (Fall 1982).

Matthews, Samuel W. "New World of the Ocean." *National Geographic* 160, No. 6 (December 1981): 792-832.

Mottl, Michael J. "Submarine Hydrothermal Ore Deposits." *Oceanus* 23, No. 2 (Summer 1980): 18-27.

Revelle, Roger. "Carbon Dioxide and World Climate." *Scientific American* 247, No. 2 (August 1982): 35-43.

CHEMISTRY AND ARCHAEOLOGY

Analyzing the Secrets of the Past

Earle R. Caley

Earle R. Caley, Professor Emeritus of Chemistry at the Ohio State University, has been interested in the application of chemistry to archaeology for more than fifty years. In 1937, he organized and operated, in Athens, Greece, the first chemical laboratory at an excavation site. In 1949 he suggested that the application of chemistry to archaeology be made a special branch of applied chemistry to be known as archaeological chemistry. This suggestion has since been generally followed. Professor Caley is the author or co-author of ten books, contributions to some twenty others, and many articles in a wide range of scholarly periodicals. He has received various honors in recognition of his work including the Lewis Prize of the American Philosophical Society (1940), a citation from the American Classical League (1954), and the Dexter Award in the History of Chemistry (1966).

Soldiers' shields, coins, pottery, toilet articles — the discovery and excavation of such articles left behind by early peoples and past civilizations is generally considered the work of archaeologists. But chemists now often cooperate in the study of excavated remains, although they approach this study with a different point of view. Archaeologists are mainly interested in the external aspects of such remains, whereas chemists are primarily concerned with their internal composition. Complete information about the age, origin, composition and purpose of ancient materials and objects is frequently possible only by a union of both points of view.

EXPERIMENTAL METHODS

Chemical analysis of the composition is the most commonly used technique for the study of ancient remains in the laboratory, though certain physical methods of measurement are now being used with increasing frequency. Methods such as X-ray fluorescence analysis which cause only slight loss of material from specimens or damage to objects are usually preferred. Used for determining the composition of an ancient metallic object, for example, this technique directs a beam of X-rays towards a small cleaned flat area on the surface of the object; the resulting reflected radiation, measured with a suitable detector, indicates what metals are present and what their percentages are. Still more preferred are the very few methods that do not cause any appreciable loss of material or damage to objects. In neutron activation analysis, for example, the entire object, a coin, for example, is bombarded with neutrons (high-speed subatomic particles lacking an electrical charge), which cause a small proportion of the atoms in the coin to become radioactive. Measurement of the emitted beta and gamma rays shows the identity and proportion of most of the metals present. The appearance of the coin is not changed by this treatment, and the induced radioactivity drops to an insignificant amount after a few days.

IDENTIFICATION OF MATERIALS

The identification of ancient materials on the basis of superficial appearance rather than chemical tests often leads to erroneous conclusions. For example, some white metal ornaments found in Mexico were first thought to be silver objects, but tests showed that they were composed of almost pure tin. Objects made of silver containing small proportions of copper cannot be distinguished by color from pure silver objects, but such a distinction may be significant when the objects are ancient coins. It is essential to determine whether a metal object found at a Mediterranean site is composed of brass (a copper-zinc alloy) or of bronze (a copper-tin alloy), because identification as the former would indicate a date of manufacture in Roman times, whereas identification as the latter could indicate a date centuries earlier.

Chemical tests allow accurate conclusions as to the identity of the original material of a completely corroded ancient object. For example, an object composed of green corrosion products found in an excavation in Athens had the shape of a soldier's shield. It was first believed that the shield had been made of bronze, but the lack of tin compounds in the corrosion products proved that this belief was incorrect. Chemical tests revealed that the shield had been made of almost pure copper.

Chemical tests are also a necessity when visual inspection of ancient materials yields no clues as to their identity. For example, lumps and powder of a white substance were found in ornamental ceramic containers in graves of women buried in the third and fourth centuries B.C. in southern Greece. These graves also contained various toilet articles. After some speculation as to whether this material was talc, chalk, white clay, or some mixture of these, chemical analyses showed that it was white lead (manufactured basic lead carbonate), evidently used as a cosmetic powder. The identification of the cosmetic material as white lead dates the discovery and use of a process for the manufacture of this pigment, its sale as an article of commerce, and the popular use of a poisonous substance in a way that may well have had an adverse effect on the public health.

The chemical examination of ancient materials may indicate their actual source. For example, deposits of copper do not occur in Ohio, yet large numbers of prehistoric copper objects have been found in that state. The copper for the production of these objects or the objects themselves must obviously have been imported. Because the copper deposits of the Lake Superior region are the nearest, they seemed to be the most likely source. That this was indeed true was demonstrated by the essential identity of samples of metal taken from the objects and those from Lake Superior copper. Similarly, the results of chemical tests made on ancient amber objects from sites in southern Greece showed that the amber of these objects was the kind found only on the shores of the Baltic Sea. Such determinations of distant sources of raw materials or finished objects provide evidence of ancient trade relations.

COMPOSITION OF COINS

In classical Mediterranean civilizations the weight and purity of precious metal in essentially silver coins were excellent reflections of the economy of the civilization. Decrease in the purity of the silver was a debasement of the coinage indicative of adverse economic conditions. Chemical analyses of many silver coins of the Roman emperors show a general trend towards severe debasement over a period of three centuries. Thus, during the reign of Augustus (27 B.C. - A.D. 14), at the beginning of the empire, the average percentage of silver in the principal silver coin, called the denarius, was about 99%. The average percentage declined slowly in the first century of the empire, more rapidly in the second century, and still more rapidly in the third century, until by A.D. 275 the principal coins contained only about 5% silver, most of it on the surface in the form of plating. Analyses also show that a similar debasement of silver coins occurred in colonies of the empire, especially in Egypt.

The results of chemical analyses also date the introduction of new alloys. For example, Greek coins of the third century B.C. were made of bronze containing considerable lead, but brass as a coinage alloy was introduced by the Romans in the first century B.C.

AGE DETERMINATION

Some chemical and radiochemical methods are available for estimating the age of ancient materials and objects. The earliest and most widely used method is applicable to materials containing carbon that came from living organisms. This method depends on the natural occurrence of weakly radioactive carbon of atomic weight 14 (carbon-14) in small proportion in the carbon dioxide (CO_2) of the atmosphere. When carbon dioxide is utilized by a growing tree to form wood, the carbon in the wood contains radioactive carbon in the same proportion as was present in the carbon dioxide of the atmosphere at that time. As soon as the wood is harvested, the radioactivity of the carbon in it starts to dwindle spontaneously. Carbon-14 has a half-life of 5800 years, i.e., the rate of decrease is such that the radioactivity has decreased to half its original intensity after 5800 years and to half that again after another 5800 years. Hence the age of a piece of ancient wood can be estimated by isolating some of the carbon contained in it, measuring the present radioactivity of this carbon, and making a simple calculation. The age of wooden objects and other plant materials may be estimated in this way. Because of past erratic fluctuations in the radioactivity of atmospheric carbon, ages estimated by this method may be in slight error. Sometimes a correction can be made by means of data obtained from radiocarbon measurements of the age of wood taken from tree rings of precisely known date.

ANCIENT TECHNIQUES

Information about ancient manufacturing formulas or processes may sometimes be obtained from the chemical investigation of excavated materials and objects. For example, some ancient pottery glazes from Tarsus (in what is now southern

FIGURE 1: Excavation Finds, Receiving Department—Agora Excavations, Athens, Greece, 1937

Turkey), which were investigated because of their unusual brilliance, were found to contain high proportions of lead. This was very probably added, in the form of lead oxide, to the glaze mixture before firing. Some dark blue glazes were found to owe their unusual color to the addition of cobalt compounds.

Experiments imitating ancient technical processes can yield information about details of operations. A considerable number of these experiments have been made on the reproduction of pottery bodies and glazes. It has been shown, for example, that the black glaze on ancient Greek pottery of the classical period could only have been produced by firing in a kiln that contained little or no oxygen.

Although the uses to which ancient objects were put are often obvious from simple inspection, a chemical investigation may sometimes be helpful for determining the exact use of a given object or class of object. This is especially true for containers. For example, in excavations at the ancient Greek marketplace in Athens large numbers of fragments of empty jars were found. These were coated on the inside surface with a black or dark brown resinous material. Since the unglazed pottery body was porous, it was evident that the purpose of the coating was to render the jars impervious to liquids. This coating was found to be soluble in vegetable oils but not appreciably soluble in alcoholic solutions containing less than 20% alcohol. Hence the jars could not have held oils such as olive oil but could have been used to hold wines. Stamped impressions on their handles showed that the jars were of foreign origin. Therefore it is probable that these jars were containers for imported wine. Obviously they were not originally containers for water, though they may have been used for this purpose later.

Tests on a number of samples of the coating showed that they all consisted of mastic resin that had been altered by fusion. The natural resin occurs as an exudation from a shrub or small tree native to the eastern Mediterranean region.

ALTERATIONS IN APPEARANCE

A great variety of chemical reactions occur between a material buried in the ground for a long time and the surrounding soil and ground water. In order to explain completely the appearance of objects found in excavations, to determine with confidence their probable original appearance, and to decide on the proper method of restoration, it is often essential to understand the nature and course of these chemical reactions. Objects of the same material and of the same age taken from various excavations may differ in appearance and condition. A bronze from one site may be coated with a hard coherent layer of corrosion products of the sort called a patina, whereas a similar bronze of the same age from another site may be coated with a loose porous mass of corrosion products. Archaeologists observed that bronze found at Corinth in Greece was almost invariably in a severely corroded state, whereas similar bronze of equal or greater age found at most other sites in Greece was much less corroded. An investigation into the reason for this difference showed that the ground water at Corinth contained an abnormally high concentration of sodium chloride and other chloride salts. This accounted for the severe corrosion of the buried bronze.

RESTORATION AND PRESERVATION

Since the deterioration of buried objects is largely the result of chemical changes, most such objects can be restored by chemical treatment. With certain kinds of materials such as bronzes, the usual restoration process is essentially a reversal of the chemical changes that produced the deterioration. In the corrosion of a bronze, the surface metal is changed to compounds of the various metals contained in it. These changes are initiated and promoted by weak electrical currents that pass between points of different composition or structure on the moist surface of the bronze. If one applies an electrical current to the corroded bronze object in the proper way, the compounds on the surface will change back to the original metals. The simplest procedure for doing this is to imbed the corroded bronze in granulated zinc in a glass container and fill the vessel with a 5% sodium hydroxide ($NaOH$) solution. After a few days the bronze object is removed, scoured with a wire brush, washed repeatedly with distilled water, and dried in an oven.

Even when objects are of such a nature that they undergo no appreciable chemical deterioration during burial, they may become coated with foreign matter that requires chemical treatment for effective and safe removal. For example, an inscribed cone of baked clay found in Sumeria could only be partially deciphered because most of it was coated with an adherent, opaque layer of calcium carbonate ($CaCO_3$) and calcium sulfate ($CaSO_4$) deposited by ground water. This layer could not be removed mechanically without flaking off the softer inscribed surface of the cone. However, treatment with dilute hydrochloric acid (HCl), followed by washing and drying, removed all the deposit safely so that the entire inscription could be read.

Whether ancient objects are restored by chemical means or not, problems concerning their preservation for exhibition or study often arise, and these problems are frequently best solved by the application of chemical principles. The corrosion of an ancient metal object, for example, may continue in ordinary air at an appreciable rate unless proper precautions are taken. Such an object may sometimes be preserved by coating it with a protective layer of wax or synthetic resin. When such treatment is

FIGURE 2: Temporary Field Museum of Excavated and Restored Objects—Agora Excavation Project, Athens, Greece, 1937

not desirable, the object may be kept in a sealed glass case filled with an inert gas or in a case containing a powerful drying agent.

In recent years the volume of restoration and conservation work in the larger museums has increased to such an extent that many of them have established their own laboratories. In this country the Boston Museum of Fine Arts, for example, has such a laboratory, and in England the British Museum maintains a large laboratory housed in a separate building.

We have seen that chemistry can be of service to archaeology in a variety of ways, and additional examples can be found in the suggested readings. Some of these references also contain detailed procedures for making the above-mentioned tests. Also, some of the suggested readings will be helpful in locating many original articles on the application of chemistry to archaeology.

SUGGESTED READINGS

BOOKS

Brill, R. H. *Science and Archaeology*. Cambridge, Mass.: MIT Press, 1971.

Caley, E. R. *Analyses of Ancient Glasses*. Corning, N.Y.: The Corning Museum of Glass, 1962.

Caley, E. R. *Analysis of Ancient Metals*. New York: The Macmillan Company, 1964.

Carter, G. F. *Archaeological Chemistry II*. Washington: American Chemical Society, Advances in Chemistry Series No. 171, 1978.

Levey, Martin (ed.). *Archaeological Chemistry, A Symposium*. Philadelphia: University of Pennsylvania Press, 1967.

Libby, W. F. *Radiocarbon Dating*. Chicago: University of Chicago Press, 1952.

Plenderleith, H. J. *The Conservation of Antiquities and Works of Art*. New York: Oxford University Press, 1956.

Rosenfeld, Andrée. *The Inorganic Raw Materials of Antiquity*. New York: Frederick A. Praeger Publishers, 1965.

ARTICLE

Zurer, Pamela S. "Archaeological Chemistry: Physical Science Helps to Unravel Human History." *Chemical and Engineering News* 61, No. 8 (February 21, 1983): 26-44.

CHEMISTRY AND WARFARE

From Destruction to Mankind's Health and Benefit

George B. Kauffman

George B. Kauffman, Professor of Chemistry at California State University, Fresno, is the author or editor of 11 books and more than 500 technical papers, reviews, and encyclopedia articles on chemistry, chemical education, and the history of science. A former Guggenheim Fellow (1972), Dr. Kauffman is the recipient of many awards and honors, among them the Dexter Award in the History of Chemistry (1978), the Outstanding Professor Award of the California State University and Colleges System (1973), the Manufacturing Chemists Association Award for Excellence in College Chemistry Teaching (1976), and the Lev Aleksandrovich Chugaev Jubilee Diploma and Bronze Medal of the U.S.S.R. Academy of Sciences (1976). He has been Contributing Editor of the Journal of College Science Teaching *since 1973, Editor of the American Chemical Society's History of Chemistry Audiotape series (1975-1981), Contributing Editor of the* Hexagon *since 1980, and Contributing Editor ("Historical Sketches" feature) of* Polyhedron *since 1982. In 1982 he was awarded grants from the National Endowment for the Humanities and Svenska Institutet (the Swedish Institute) for his research project "A Humanist Genius as Amateur Scientist: August Strindberg's Chemical and Alchemical Studies and Their Influence on His Literary and Dramatic Productions."*

Man has always been the most destructive as well as the most ingenious of all living creatures. Chemical research, both pure and applied, has made multifaceted contributions to both aspects of man through its contribution to warfare, both ancient and modern. The interaction between chemistry and warfare has been and continues to be a "two-way street"; not only has chemistry contributed to the technology of destruction, but the exigencies and demands of war have resulted in a host of scientific and technological "spin-offs" with significant peacetime uses.

Primitive man first fought with his bare hands, feet, and teeth, but as early as the middle Pleistocene Epoch (about 250,000 years ago) he was already using crude arms such as wooden or stone clubs. Spears and axes could be used to strike enemies before they reached grappling range. The search for more effective weapons led, over the course of thousands of years, to the replacement of stone weapons by those made of metal. In fact, iron was first used in implements of battle. The sword preceded hoes, picks, and other tools.

ARCHAEOLOGICAL AGES AND THE ACTIVITY SERIES

The advance of weaponry and of civilization itself can thus be correlated with the increasing skill of primitive man in the art of metallurgy, the technique of extracting metals from their ores (minerals in which they occur), which is a branch of applied chemistry. The practice of metallurgy is intimately connected with the relative activity of metals, which is best shown in the familiar activity series known to every student of high school chemistry (Table 1). In this series the metallic elements are arranged in order of decreasing activity.

The most active metals, located near the top of the table, are very reactive and therefore never occur in nature as free metal but only in the form of compounds from which they can be isolated only with considerable difficulty. The least active metals, located near the bottom of the series, are unreactive and therefore often occur in nature as free metal. Thus the advance of metallurgical technology corresponds to the ability of man to extract more and more active metals from their ores, that is, to isolate metals that appear higher and higher in the activity series.

Prehistoric archaeological time can be divided into periods according to the materials of which man's implements and weapons were fashioned. Thus the Stone Age (subdivided into Lower, Middle, and Upper Paleolithic, Mesolithic, and Neolithic) was followed by the Bronze Age and then the Iron Age. During the Stone Age mankind's knowledge of metallurgy was nil; during the Bronze Age man acquired the skill to smelt ores of copper and tin to produce bronze (an alloy of copper and tin). With the advent of the Iron Age he was able to extract the more active metal, iron, from its ore, a more difficult task than extracting the less active metals, copper and tin, from their ores.

Exact dates for the archaeological ages cannot be specified, for different regions of the earth passed through these ages at different times. Even today, we occasionally read of the discovery of a primitive tribe still living in the Stone Age. Obviously, nations or tribes with greater degrees of expertise in metallurgy possessed distinct military advantages over their less advanced neighbors. Thus the Philistines, a cultured, nonsemitic race which entered Canaan from the Mediterranean at about the same time as the Hebrews entered from the desert, were familiar with the technology of iron and were in their Iron Age, while the Israelites were still in their Bronze Age. The Philistines wisely retained their monopoly on working the metal, for in *I Samuel,* 13: 19-22 we read:

> Now there was no smith found throughout all the land of Israel: for the Philistines said, Lest the Hebrews make them swords or spears; But all the Israelites went down to the Philistines, to sharpen every man his share, and his coulter, and his ax, and his mattock. Yet they had a file for the mattocks, and for the coulters, and for the forks, and for the axes, and to sharpen the goads. So it came to pass in the day of battle, that there was neither sword nor spear found in the hand of any of the people that were with Saul and Jonathan: but with Saul and with Jonathan his son was there found.

The end of Philistine domination also ended the embargo on iron, and by the time that David ascended the throne in about 1000 B.C., the use of iron by the Israelites was more general. By the time of the prophet Amos (about 760 B.C.) its use was quite common, and in *Jeremiah*, 11:4, written about 600 B.C., we read:

> Which I commanded your fathers in the day that I brought them forth out of the land of Egypt, from the iron furnace.

Activity	Element	Discovery Date	Reduction of Oxides
Very active with water or acids	Potassium, K	1807	Oxides reduced by electrolysis but not by hydrogen or carbon monoxide
	Barium, Ba	1808	
	Calcium, Ca	1808	
	Sodium, Na	1807	
Active with acids or with steam when metal is hot	Magnesium, Mg	1808	
	Aluminum, Al	1827	
	Manganese, Mn	1774	Oxides reduced by carbon or aluminum (hot) but not by hydrogen or carbon monoxide
	Zinc, Zn	Known to Ancients as Brass	
	Chromium, Cr	1797	
	Iron, Fe	Known to Ancients	
Less reactive with acids	Cadmium, Cd	1817	
	Nickel, Ni	1751	

Reactivity with acids	Element	Discovery	Reduction of oxides
	Tin, Sn	Known to Ancients	
	Lead, Pb	Known to Ancients	
	Hydrogen, H	16th-17th Century	Oxides reduced by heating with hydrogen or carbon monoxide
	Copper, Cu	Known to Ancients	
Reactive only with oxidizing acids such as nitric acid (HNO_3)	Antimony, Sb	Known to Alchemists, Possibly to Ancients	
	Bismuth, Bi	Known to Alchemists	
	Mercury, Hg	Known to Ancients	
	Silver, Ag	Known to Ancients	
Nonreactive with single acids; react with *aqua regia* (hydrochloric acid: nitric acid = 4:1)	Gold, Au	Known to Ancients	Oxides reduced to metal by heat alone
	Platinum, Pt	1748	

TABLE I: Activity Series of Some Metallic Elements

In an extension of the archaeological time periods, the twentieth century has been called "The Age of Steel," a name given by the Russian composer Serge Prokofiev to his ballet of 1927, "Le Pas d'Acier". The strength and military importance of this alloy of iron is also attested to by the pseudonym chosen by Josif Vissarionovich Djugashvili, better known to us as Joseph Stalin (1879-1953). Stalin comes from the Russian word *stal* for steel. Because of the importance of nuclear energy, our present time, beginning with 1945, has been christened by some "The Atomic Age," a term that we shall refer to again in more detail.

Of the seven metals known to the ancients (gold, silver, iron, mercury, tin, copper, and lead), four (gold, silver, mercury, and copper) are below hydrogen in the activity series; the more active metals have been isolated only comparatively recently. Thus the alkali metals (periodic group IA — lithium, sodium, potassium, rubidium, and cesium) and alkaline earth metals (periodic group IIA — beryllium, magnesium, calcium, strontium, and barium) were not isolated until after the Italian physicist Count Alessandro Volta (1745-1827) had invented the so-called voltaic pile in about 1800. With the aid of this source of electrical potential, Sir Humphry Davy (1778-1829) isolated the active alkali metals potassium and sodium (both in early October, 1807), followed in short order by the alkaline earth metals barium, strontium, calcium, and magnesium in 1808.

Aluminum is an extremely important metal with numerous military uses. The inspiring story of the development of the process for its isolation shows how a young but enthusiastic student can make a scientific discovery of tremendous significance. Aluminum is a light, very active metal, which can be used structurally in the aircraft industry only because it acquires a protective, adherent, thin layer of aluminum oxide. It is the most abundant metal on the earth's surface, comprising about 8% of the earth's crust, a percentage exceeded only by oxygen and silicon, and the most widely used metal with the exception of iron and steel. Although its crystalline double sulfate with potassium sulfate — alum — was a commercial article of great antiquity, being known to the ancient Greeks and Romans, the metallic element remained unknown until relatively recent times. It was not until 1827 that the young German chemist Friedrich Wöhler (1800-1882) isolated the impure metal by reducing anhydrous (water-free) aluminum chloride with potassium, a highly active metal that was discovered only twenty years earlier. It remained an extremely expensive laboratory curiosity, more expensive than

FIGURE 1: **Sir Humphry Davy, 1778-1829.** English chemist and physicist. One of the founders of electrochemistry. Inventor of the safety lamp for miners. He was the first to isolate potassium, sodium, calcium, barium, strontium, and magnesium. Davy in England and Gay-Lussac and Thenard in France, working independently, were the first to isolate boron.

Photo reproduced with permission of the *Journal of Chemical Education* from Weeks, Mary Elvira, "Discovery of The Elements," *Journal of Chemical Education*, 7th ed., (1968), p. 432.

gold. The first authentic article of aluminum was a rattle for the baby who later became the French emperor Napoleon III and who was later presented with an aluminum medal in 1854. Napoleon III authorized and financed experiments to produce me-

FIGURE 2: **The Aluminum "Crown Jewels."** In this chest, carefully preserved by the Aluminum Company of America at Pittsburgh, are the original buttons of the metal made by Charles M. Hall in Oberlin, February 23, 1886 (left), the larger ones made by Hall in December, 1886 (center), and the first button or ingot (right) produced by the Aluminum Company of America.

Photo reproduced with permission of the *Journal of Chemical Education* from Weeks, Mary Elvira, "Discovery of the Elements," *Journal of Chemical Education*, 7th ed. (1968), p. 574.

tallic aluminum on a large scale, hoping to use it for breastplates and helmets for his soldiers, but its high price ($204 per lb.) made the project a financial fiasco.

More than three decades later, Charles Martin Hall (1863-1914), a 21-year-old student at Oberlin College in Oberlin, Ohio, was inspired to search for a cheap method of producing aluminum. His chemistry professor, Frank Fanning Jewett (1844-1926), had worked under Wöhler and told his class of Wöhler's work on aluminum. On February 23, 1886 Hall produced the first buttons of aluminum made by electrolysis with the use of homemade batteries. He worked in an improvised woodshed laboratory. The so-called Hall process was a great commercial success, and between the years 1886 and 1890 the price of aluminum had dropped from $3 per lb. to 60¢ per lb. By the year 1914, when Hall died at the early age of 51 leaving a fortune of nearly $30 million, technology had advanced to the point where the price of aluminum had dropped to 18¢ per lb. During World War I aluminum alloys were employed by Germany in the framework of Zeppelin airships and planes. By 1939 bulletproof armor made of duralumin (an aluminum alloy containing up to 5% copper and small amounts of magnesium, manganese, silicon, and iron) was used extensively by German forces on the Western Front. By 1941 the world consumption of aluminum, which had been a mere laboratory curiosity only a half century earlier, was close to a million tons.

DYES IN WARFARE

Compounds used as dyes and pigments stem from earliest antiquity. For example, the central event around which the Old Testament revolves is undoubtedly the Exodus of the Israelites from Egypt. As everyone who has ever attended Sunday school knows, the Israelites were descendants of Jacob and his sons, who settled in Egypt at the behest of their son and brother, Joseph, who was the Pharaoh's second-in-command. Joseph had been sold into Egypt by Midianites after being placed in a pit by his brothers, whose sibling rivalry had been aroused by his dreams of glory and his father's blatant preference for him symbolized by "a coat of many colors" (*Genesis,* 37:3). Thus Joseph's coat, probably dyed with an alizarin dye, played an important and possibly pivotal role in the central event of Biblical history.

The bright red alizarin dye was involved in what was probably one of the earliest examples of chemical camouflage. A calcium aluminum chelate compound of hydroxyanthraquinone, alizarin was first used in India and was known to the ancient Persians and Egyptians long before it was used by the Greeks and Romans. Once, after a day of battle with a much larger Persian force, Alexander the Great had his soldiers' uniforms dyed with bloodlike splotches using alizarin. On the dawn of the second day, the Persians, thinking that many of the Macedonians had been wounded during the previous day's fighting and had not received medical care, heedlessly attacked without proper caution and were soundly defeated by Alexander's smaller army.

A more recent example of the effect of this dye on the course of history can be found in our own Revolutionary War. Anyone who has seen any Hollywood movies dealing with this period knows that George III's troops marched with military precision in a straight line, while the American revolutionaries shot at them from behind rocks and trees. To add to our advantage, the British soldiers made even better targets because their uniforms were dyed with madder and clay, which gave them the hated nickname of "redcoats." To the colonists, this uniform soon became the symbol of British oppression.

"GREEK-FIRE"

"Greek-fire" — incendiary mixtures of various compositions used in warfare in medieval times by Byzantine Greeks — probably dates from ancient times. The Roman historian Livy (59 B.C.-A.D. 17) wrote of the Bacchantes (priests, priestesses, or votaries of Bacchus, the Greek god of wine) carrying torches containing sulfur and quicklime (calcium oxide, CaO) which inflamed on dipping in water. Ingredients common to Greek fire mixtures included sulfur, bitumen (tar, pitch, asphalt, or petroleum), *sal petrosum* (potassium nitrate, KNO_3), and quicklime. Sulfur and bitumen were used not only because they burned well but because they adhered to targets while burning (the ancient and medieval equivalents of napalm). *Sal petrosum*, which liberates oxygen on heating, caused mixtures to burn quickly and vigorously and also made them difficult to extinguish. Quicklime was used for mixtures to be ignited by dampening or immersion in water because of its great heat of hydration. The *Liber Ignium ad Comburendos Hostes* (Book of Fires for Burning Enemies), attributed to a certain Marcus Graecus and apparently dating from the late thirteenth century, describes an oily mixture to be placed in a goat-skin bottle and floated on a piece of burning wood into the midst of an enemy fleet. The bottle is soon burnt through, and the flaming mixture quickly spreads across the water.

GUNPOWDER

Gunpowder or black powder, a mixture of saltpeter or *sal petrosum*, charcoal, and sulfur had a more revolutionary effect on warfare. It was the first and only explosive known until the discovery of fulminating gold, a powerful explosive first used in European wars in 1628. In his classic book, *The Decline and Fall of the Roman Empire* (1776-1788), the English historian Edward Gibbon (1737-1794) wrote:

> The military art has been changed by the invention of gunpowder, which enables man to command the two most powerful agents of nature, air and fire . . . If we contrast the rapid progress of this mischievous discovery with the slow and laborious advances of reason, science and the arts of peace, a philosopher, according to his temper, will laugh or weep at the folly of mankind.

This earliest of explosives was invented during the Southern Sung Dynasty (1127-1279) by the Chinese, who used it for fireworks and military purposes. The Mongols probably introduced it into Europe in about 1241. The English friar, Roger Bacon (1214?-1294?), described its preparation, and a German monk, Berthold Anklitzen, better known as Berthold Schwarz (Berthold the Black), is said to have invented firearms by using gunpowder to propel a projectile in 1313. Small crude cannons were first used by the English against the French in the Battle of Crécy in 1346, the first major encounter of the Hundred Years' War. Although firearms were not decisive in the outcome of this war, improved weapons were soon developed, and they eventually contributed to the decline of knighthood, walled cities, and the entire feudal system of medieval Europe.

MODERN EXPLOSIVES

The first modern explosives were nitrocellulose and nitroglycerin. Nitrocellulose (guncotton) was discovered in 1845 by Christian Friedrich Schönbein (1799-1868), a German chemist, who, while experimenting in his wife's kitchen with a mixture of nitric and sulfuric acids, accidentally spilled some of the mixture. Since his wife had forbidden him to use her kitchen for his work, he cleaned up the mixture with the first thing that he could find — his wife's cotton apron, which he then washed, wrung out, and hung over the stove to dry. The apron not only dried but also was consumed in a sudden puff of flame, and Schönbein, recognizing the significance of his discovery, went on to establish factories to manufacture smokeless powder from guncotton. Schoñbein kept his method of preparation secret, but secrecy in science and technology is difficult to maintain, as the United States later found with the atomic bomb after World War II. In 1846, the same year that Schönbein announced the discovery of guncotton, it was discovered independently by German chemists Rudolph Christian Böttger (1806-1881) and Friedrich Julius Otto (1809-1870). Like many discoveries that have destructive uses, guncotton has had peaceful applications as well. Schönbein found that guncotton could be dissolved in ether, and the resulting collodion was used in medicine to coat wounds, in photography for the manufacture of film, and in industry for lacquers and plastics.

Nitroglycerin (glyceryl trinitrate) was discovered in 1846 by the Italian chemist Ascanio Sobrero (1812-1888), who added glycerin to a mixture of concentrated nitric and sulfuric acids. It has a high nitrogen content and contains more than enough oxygen to oxidize the carbon and hydrogen that it contains. On detonation, nitrogen gas is liberated so that it is one of the most powerful explosives known. Its detonation generates gases that occupy more than 1200 times the original volume at ordinary room temperature and pressure, while the heat liberated raises the temperature to about 5,000°C. (9,032° F.). The overall effect is the instantaneous development of a pressure of 20,000 atmospheres (294,000 pounds per square inch). Nitroglycerin is so sensitive to shock that it was not used as an explosive until 1866 when the Swedish engineer, inventor, philanthropist, and originator of the Nobel Prizes, Alfred Bernhard Nobel (1833-1896), mixed it with such inert adsorbents as diatomaceous earth to form dynamite.

Nitroglycerin converts nitrocellulose (guncotton) to blasting gelatin, a powerful explosive. Nobel's discovery of this gelatinizing action led to the development of ballistite, the first double-base propellant and a precursor of cordite, a smokeless, slow-burning powder. Nitroglycerin also has peaceful uses, being widely used in medicine to dilate the veins and thus relieve the pain (angina pectoris) in heart ailments.

During World War I, 2,4,6-trinitrotoluene, commonly known as TNT, was the most generally employed high explosive, and it is still the most useful militarily. Before and during World War II a number of extremely efficient new high explosives were developed, including cyclonite or RDX and pentaerythritol nitrate or PETN.

POISON GASES

World War I also witnessed the first large-scale poison gas attack in military history. In a bold attempt to break the deadlock of static trench warfare on the Western Front, the Germans, late on the afternoon of April 22, 1915, used chlorine gas to penetrate the Allied trenches near Ypres in Belgium, where the British and French lines met. The Allies were caught unprepared, but the Germans failed to foresee the success of their surprise attack and did not fully exploit their advantage. On September 25 and 27, 1915 the British retaliated with their own chlorine attack at Loos, also in Belgium. The Germans introduced phosgene (carbonyl chloride, $COCl_2$) in December of 1915 and mustard gas (2,2′-dichlorodiethyl sulfide) in July of 1917.

In 1917 the United States government built the Edgewood Arsenal on a 10,000-acre site in Maryland and installed a chemical warfare school. A huge chlorine plant — probably the largest in the world at that time — capable of filling more than

100,000 shells and bombs per day was built. About 1,200 technical persons and 700 assistants worked at Edgewood. The U.S. chemical warfare effort was probably the largest research organization devoted to a single goal up to that time.

On June 28, 1918 the United States War Department consolidated chemical warfare activities by creating the Chemical Warfare Service. By the end of the war, the United States was manufacturing nearly four times the amount of poison gas produced by the Germans. During the war, no less than 28 gases and 16 mixtures of gases were used by the Allies and the Central Powers. Before the war, the United States and other countries were largely dependent upon Germany for dyes and heavy and fine chemicals. A by-product of the war was the emergence of the American chemical industry, which has made this country not only self-sufficient but a leader in this field. The universal revulsion against gas warfare, along with its indecisive effect, produced a virtual outlawing of it. Thus poison gases were not used in World War II, although both the Allies and the Axis powers produced gases and gas masks in quantity during the conflict.

NUCLEAR WEAPONS

The best kept secret of World War II and the development that had the farthest reaching consequences was without doubt the atomic bomb (really a nuclear bomb since nuclear, rather than atomic, reactions are involved). Actually, the events leading to the development of this "super-weapon" began with the discovery of radioactivity by the Frenchman Antoine Henri Becquerel (1852-1908) in 1896. Brilliant discoveries followed by Marie Curie (1867-1934), Pierre Curie (1859-1906), Ernest Rutherford (1871-1937), William Crookes (1832-1919), Frederick Soddy (1877-1956), Otto Hahn (1879-1968), Kasimir Fajans (1887-1975), Alexander S. Russell (1888-1972), Frédéric (1900-1958) and Irène Joliot-Curie (1897-1956), and others. In 1939 Otto Hahn and Fritz Strassmann (1902-1980) in Germany and Lise Meitner (1878-1968) in Sweden discovered the nuclear fission of uranium, a contribution for which they shared the 1966 Enrico Fermi Award of the United States Atomic Energy Commission. From this point on, work began in earnest on an atomic bomb. During World War II the United States launched a $2 billion effort, code-named the Manhattan Project, which employed 125,000 persons, of whom about 10,000 were scientists. On December 2, 1942 the first self-sustaining nuclear chain reaction was carried out under Stagg Field at the University of Chicago. Here Enrico Fermi (1901-1954) and his co-workers assembled tons of graphite, uranium, and uranium oxide into the first nuclear reactor. The experiment proved that the release of nuclear energy could be controlled and thus ushered in the Atomic Age.

In order to separate the fissionable isotope uranium-235 (the form of uranium having an atomic weight of 235), which occurs in natural uranium to the extent of only 0.72% (See Table II), a huge and expensive gaseous diffusion plant was built in the backwoods of Tennessee near the town of Oak Ridge. Uranium was converted into the gaseous compound uranium hexafluoride (UF_6), which was allowed to diffuse through porous barriers. After numerous repetitive diffusion cycles, the UF_6 containing the fissionable U-235 was separated from the UF_6 containing the nonfissionable U-238 and U-234.

Isotope	*% Abundance*	*Protons*	*Neutrons*	*Electrons*	*Half Life (Years)*
U-238	99.27	92	146	92	4,510,000,000
U-235	0.72	92	143	92	713,000,000
U-234	0.006	92	142	92	248,000

TABLE II: The Isotopes of Uranium (Atomic Number 92)

Since the success of this project could not be assured, three enormous reactors were constructed at a cost of $50 million on the banks of the Columbia River at the General Electric Hanford plant in the State of Washington in order to produce an alternative bomb material, the transuranium element plutonium (Pu, atomic number 94), which had been discovered at the University of California, Berkeley by Glenn Theodore Seaborg (born 1912), Arthur Charles Wahl (born 1917), and Joseph William Kennedy (1916-1957) on December 14, 1940. Because of security considerations, the report of the discovery of plutonium was not published until 1946. On March 28, 1941 Seaborg and co-

FIGURE 3: **Enrico Fermi, 1901-1954.** Naturalized Italian-American physicist. Professor of physics at Columbia University and the University of Chicago. In 1938 he was awarded the Nobel Prize for Physics in recognition of his work on artificial radioactivity induced by bombardment with neutrons. He found that the effectiveness of neutron bombardment is much greater in the presence of water or paraffin and concluded that the neutrons are slowed down by collisions with the hydrogen nuclei in these substances and therefore have a greater probability of disrupting nuclei. The citation accompanying his Congressional Medal for Merit, awarded in 1946, states that he was "the first man in all the world to achieve nuclear chain reaction."

Photo reproduced by Argonne National Laboratory and reprinted with permission of the *Journal of Chemical Education* from Weeks, Mary Elvira, "Discovery of the Elements," *Journal of Chemical Education*, 7th ed. (1968), p. 833.

workers showed that plutonium-239 undergoes fission by slow neutrons.

On July 16, 1945, at 5:30 A.M., the first atomic bomb was detonated at Trinity Site, Alamogordo Air Force Base, New Mexico, about 120 miles southeast of Albuquerque. Less than a month later a U-235 atomic bomb (code-named "Little Boy") was dropped from the bomber "Enola Gay" on the city of Hiroshima, Japan. The resulting blast, with a force equivalent to 20,000 tons of TNT, killed about 70,000 persons. Three days later, on August 9, 1945, a plutonium atomic bomb (code-named "Fat Man") was dropped on Nagasaki, Japan, resulting in 36,000 deaths. Japan quickly sued for peace, bringing the war to an end.

FIGURE 4: E. O. Lawrence, G. T. Seaborg, and J. R. Oppenheimer at Controls of Cyclotron

Photo reproduced by Argonne National Laboratory and reprinted with permission of the *Journal of Chemical Education* from Weeks, Mary Elvira, "Discovery of The Elements," *Journal of Chemical Education*, 7th ed. (1968), p. 830.

Although the dropping of the bombs is still controversial in some circles, it is said to have avoided the necessity for invading the Japanese mainland and thus to have prevented as many as one million American casualties. At any rate, it marked the advent of nuclear weapons and the beginning of the "balance of terror" under which we now live. A mere four years later — in 1949 — the Soviet Union successfully tested its own first atomic bomb, thus ending the American monopoly in nuclear weapons. On November 1, 1952 near Eniwetok Atoll in the Pacific the United States detonated the world's first thermonuclear device — the hydrogen bomb — with a force of 14,000,000 tons of TNT. In August of 1953 the Soviet Union exploded a hydrogen bomb in the megaton range. Since then, many countries have developed and detonated their own nuclear weapons. Not all these developments have been destructive, for along with the development of nuclear weapons has come the development of nuclear energy, an alternative for a world rapidly running out of fossil fuels, as well as the use of radioisotope tracers in medicine, analytical chemistry, etc. Furthermore, the Manhattan Project led to the discovery of a hitherto unknown family of elements with atomic numbers greater than that of uranium (atomic number 92), which for many years was considered to be the heaviest element. Many of these elements (atomic numbers 93 through 109) were discovered by American scientists, particularly those working at the University of California, Berkeley.

PROCESSES AND INDUSTRIES IN WARFARE

A fairly common sort of interaction between chemistry and history is the development of processes and industries caused by the exigencies of war. A few examples should suffice to illustrate this phenomenon.

The Beet Sugar Industry

During the continental blockade by the British Navy (1805-1815), France was cut off from her West Indies source of sugar cane. Scientists and industrialists, encouraged and subsidized by the Emperor Napoleon I, turned their efforts to large-scale production of sugar from sugar beets. Influenced by earlier research by the German chemist Andreas Sigismund Marggraf

FIGURE 5: The atom bomb exploded on Able Day in the air over the fleet at Bikini. Photo from a Navy patrol bomber flying just beyond the range of the explosion. The cloud is 35 miles high. From *Dawn Over Zero: The Story of the Atomic Bomb*, by William L. Laurence (Westport, CT: Greenwood Press, 2nd ed., 1946, p. 288).

(1709-1782), a Frenchman, Louis Augustin Guillaume Bosc-Dantic (1759-1828), called for a committee to be set up for this research. This done, a group of men began work in 1806. After six years of incessant toil, Jules-Paul Benjamin Delessert (1773-1847) completed the process. On January 2, 1812 Delessert announced to the chemist Jean Antoine Chaptal (1756-1832) and to Napoleon that the process was ready for large-scale application. Delessert was awarded the Cross of the Legion of Honor on the spot; Napoleon took off his own decoration and pinned it on Delessert. By the mid-1820s, there were 250 factories for producing sugar from sugar beets in France, and the beet sugar industry is still thriving today in that country as well as in other parts of the world.

Nitrogen Fixation

World War I provided an even more important case of industrial development in the process of nitrogen fixation. Nitrogen is an important constituent of all plant and animal protein. Although we are literally living in a sea of this element (nitrogen constitutes about 78% by volume of the earth's atmosphere), the cells of most living systems, with the notable exception of certain bacteria on the root nodules of peas, beans, clover, alfalfa, and other legumes, cannot assimilate nitrogen from the air to use in synthesizing proteins. The problem, then, is to convert nonassimilable atmospheric nitrogen into nitrogen compounds that can be used by plants, a process known as nitrogen fixation.

By 1909 the German chemist Fritz Haber (1868-1934) had developed the so-called Haber process of nitrogen fixation, by which ammonia (NH_3) is prepared by the direct union of its constituent elements, nitrogen (N_2) and hydrogen (H_2):

$$N_2 + 3H_2 \rightarrow 2\ NH_3$$

The reaction is carried out at temperatures of 400-550° C. (752-1022° F.) under high pressure (100-1000 atmospheres or 1,470-14,700 pounds per square inch) and in the presence of a catalyst. The resulting ammonia can be used in preparing fertilizers or can be oxidized to nitric acid by the Ostwald Process, developed by another German chemist, Wilhelm Ostwald

(1853-1932). The nitric acid can be used in the preparation of explosives. Before the advent of the Haber and Ostwald processes, nitrates and nitric acid were obtained primarily from Chile saltpeter, a naturally occurring sodium nitrate ($NaNO_3$) found in the arid regions of Chile (nitrates are soluble and are not found in regions with appreciable rainfall). Despite the blockade by the British Navy during World War I, Germany was able to meet her need for ammonia and nitric acid for fertilizers and explosives by using the recently developed Haber and Ostwald processes. Today the Haber process has largely displaced other methods of nitrogen fixation. After Germany's defeat, Haber attempted to extract gold from seawater in order to help pay the fatherland's reparations, but in this case he was not successful.

Acetone

Another commercial process developed at about the time of World War I actually helped bring about the birth of a new nation. Chaim Weizmann (1874-1952) pursued a dual career as a chemist and as a leader of the Zionist movement. Beginning in about 1909, he added biochemical investigations to his research on dyes, and he began to study fermentation reactions in search of a strain of bacteria capable of converting carbohydrates into isoamyl alcohol — a precursor of synthetic rubber. Instead, in 1912, he discovered the bacterium *Clostridium acetobutylicum*, which breaks down starches into one part of ethanol (ethyl or grain alcohol), three parts of acetone, and six parts of butanol (butyl alcohol), and he thus opened the microbiological route to the production of industrial chemicals. During World War I, when large quantities of acetone were required to plasticize cordite, a smokeless, slow-burning propellent powder, Weizmann successfully engineered the massive production of acetone in Great Britain and became director of the British Admiralty Chemical Laboratories. Plants were also built in the United States, Canada, and India, and production continued after the war. Weizmann's wartime services to Britain were not forgotten, and in 1917 he secured from Lord Balfour the famous Balfour declaration of British assistance in establishing a national homeland for the Jews. The realization of this Zionist dream later required so much of Weizmann's time that he stopped all his scientific activity except for promoting the growth of the Hebrew University and for founding of the world-renowned institute of science at Rehovot later named the Weizmann Institute in his honor. When Israel became a state in 1948, the Knesset (Parliament) elected Weizmann first president of the new democratic republic, a position which he held until his death in 1952. Science has formed a vital, integral part of the revived, modern Jewish culture, and the former president of Israel (1973-1978), Ephraim Katzir (originally Katchalski), was a polymer chemist.

MISCELLANEOUS TECHNOLOGICAL DEVELOPMENTS OF WORLD WAR II

According to Christopher Simpson, "some of the greatest technological advances in history have sprung from the science of warfare." World War II was even more dependent on chemical technology than World War I. Government, industrial, and academic scientists exerted heroic collaborative efforts to develop new and better war supplies and to increase the production of existing materials. From this conflict emerged a whole host of new materials, many with peacetime uses.

Penicillin

The discovery of penicillin, the first of the antibiotic miracle drugs and the one that has been hailed as "the greatest contribution medical science ever made to humanity," stemmed from an accidental observation made in 1928 by the Scottish bacteriologist Alexander Fleming (1881-1955). This fortuitous observation led to his discovery of a chemical substance which would destroy infectious bacteria without destroying tissues or weakening the body's defenses.

Fleming noticed that on a petri dish of staphylococci bacteria a mold (*Penicillium notatum*), which had been introduced by accidental contamination, had dissolved the colonies of staphylococci. Unable to isolate the bactericidal substance produced by the mold, Fleming patiently continued to maintain his cultures for a dozen years. In 1939 at Oxford University, Howard W. Florey, an Australian experimental pathologist, and Ernst B. Chain, a Jewish chemist who had fled from Nazi Germany, used the relatively new technique of lyophilization (freeze-drying) to isolate penicillin from Fleming's own cultures in a completely purified form that was one million times more active than Fleming's crude substance of 1928. In contrast to Fleming, who had worked alone with micro methods, they employed no less than than six co-workers and numerous technicians, and they produced mold juice by the gallon. In 1940 they published the results of their successful treatment of infected white mice, but a completely successful test involving a human being was not accomplished until 1942 because of the limited supply of the drug.

Although penicillin was developed in an all-out synergistic effort by a combination of government, university, and industrial laboratories working with American manufacturers, physicians, and the military, much of the pioneering work was performed at the United States Department of Agriculture's Northern Regional Research Laboratory (NRRL) at Peoria, Illinois. This outstanding example of interdisciplinary achievement turned penicillin in a few short years from a drug many times scarcer and more expensive than gold (In June, 1944, its price was $10,000 per lb.) to an inexpensive but extremely potent weapon in every physician's arsenal (about $17 per lb. in 1978).

By 1943 English and American factories were producing penicillin on a large scale, and it became available for military use. By 1944 it became available for civilian use, and the following year Fleming, Florey, and Chain were jointly awarded the Nobel Prize in Medicine or Physiology. According to Robert D. Coghill, Chief of the NRRL Fermentation Division:

> Had it not been for the war, penicillin would probably not be the household word it is today. It was the Germans' fire-bombing of England which set Florey, Chain, and their team at Oxford University to looking for a better treatment for burns ... Early yields and recovery ... were very discouraging, and I am convinced that without our wartime 85% excess profits tax, enabling industry to carry on research and development with 15¢ dollars, we never would have had penicillin at all. Penicillin is thus a more-or-less direct result of World War II. It has probably saved many

more lives and eased much more suffering than the whole war cost us and it will continue to do so for generations to come ... The penicillin battle was not won by any one group, but by a tremendous cooperative effort. University laboratories, government laboratories, and industrial laboratories all worked together on it.

Miscellaneous "Spin-Offs"

Among the literally thousands of by-products produced in response to the demands of World War II are the following: high-octane aviation gasoline produced by new alkylation and catalytic cracking processes; styrene-butadiene, neoprene, and butyl synthetic rubbers to replace the Eastern supply of the natural product, which had fallen into Japanese hands; plastics such as polyvinyl chloride films to protect military supplies from deterioration in South Pacific jungles; polyethylene, used as an electrical insulator; fluorocarbons, developed at Oak Ridge as chemically resistant gaskets able to withstand the corrosive action of UF_6 and now used in kitchen and laboratory ware; silicones, used for gaskets and high–temperature lubricants; Teflon® (polytetrafluoroethylene), originally to coat machine-gun bullet chambers, now used to coat rings in artificial heart valves; antibiotics (sulfonamides, streptomycin, etc.) and other so-called "wonder drugs" to treat wounds, burns, and diseases; dichlorodiphenyltrichloroethane (DDT) to combat insect-borne diseases such as malaria and typhus; napalm, an incendiary substance produced by thickening gasoline with an aluminum soap of naphthenic and palmitic acids, used in the bombs that devastated 15 square miles of Tokyo in the raid of March 9, 1945; and nylon, used as fabric for parachutes.

CONCLUSION

From prehistoric times to the present, chemistry has been applied to warfare. Yet such efforts, whose initial goals were usually to develop more destructive and efficient methods of killing human beings, have often led to products and processes of benefit to mankind. To quote Edward Gibbon again,

> Their [the barbarians'] gradual advances in the science of war would always be accompanied . . . with a proportionable improvement in the arts of peace and civil policy.

Therefore, until the fulfillment of the dream of the prophet Isaiah, (2:4):

> They shall beat their swords into plowshares, and their spears into pruning-hooks; nation shall not lift up sword against nation, neither shall they learn war anymore

we have the consolation of knowing from past experience that at least some of the immense amounts of time, effort, and money so profligately expended for destructive, military research will ultimately result in discoveries that will improve the length and quality of human life.

SUGGESTED READINGS

BOOKS

Harris, Robert, and Jeremy Paxman. *A Higher Form of Killing: The Secret Story of Chemical and Biological Warfare.* New York: Hill and Wang, 1982.

Holley, I. B., Jr. *Ideas and Weapons.* New Haven: Yale University Press, 1953.

Leonard, T. C. *Above the Battle: War-Making in America from Appomattox to Versailles.* New York: Oxford University Press, 1978.

Nef, John Ulric. *War and Human Progress: An Essay on the Rise of Industrial Civilization.* Cambridge: Harvard University Press, 1950.

Oman, C. W. C. *A History of the Art of War in the Middle Ages.* 2 vols. Boston: Houghton Mifflin Co., 1924.

Parr, J. G. *Man, Metals, and Modern Magic.* Cleveland: American Society for Metals, 1958.

Partington, John Riddick. *A History of Greek Fire and Gunpowder.* Cambridge, England: W. Heffer, 1960.

Sheehan, John C. *The Enchanted Ring: The Untold Story of Penicillin.* Cambridge: MIT Press, 1982.

Smith, Cyril Stanley (ed.). *Sorby Centennial Symposium on the History of Metallurgy.* New York: Gordon & Breach Science Publishers, 1963.

Tunis, Edwin. *Weapons: A Pictorial History*. Cleveland: World Publishing Co., 1954.

Weller, Jac. *Weapons and Tactics: Hastings to Berlin*. New York: St. Martin's Press, 1966.

Wertime, T. A., and J. D. Muhly (eds.). *The Coming of the Age of Iron*. New Haven: Yale University Press, 1980.

Yadin, Yigael. *The Art of Warfare in Biblical Lands in the Light of Archaeological Study*. Translated from the Hebrew by M. Pearlman. New York: McGraw-Hill Book Co., 1963.

ARTICLES

Kauffman, George B. "Chemistry and Human Affairs: Some Influences of Chemistry on History." *NEACT Journal* 5, No. 1 (Spring-Summer 1984); "Multifaceted Contributions of Chemistry to History." *Journal of College Science Teaching* 11, No. 6 (May 1982): 341-346; "The Penicillin Project: From Petri Dish to Fermentation Vat." *Chemistry* 51, No. 7 (September 1978): 11-17; "The Discovery of Penicillin — Twentieth Century Wonder Drug." *Journal of Chemical Education* 56, No. 7 (July 1979): 454-455; "Maize, Melon and Mould — Keys to Penicillin Production." *Education in Chemistry* 17, No. 6 (November 1980): 180.

Kauffman, George B., and Paul M. Priebe. "The Foundations of the Beet Sugar Industry." *Journal of Chemical Education* 56, No. 8 (August 1979): 503.

Kincaid, J. F. "Chemistry, Solid Propellants and History." *Journal of Chemical Education* 59, No. 10 (October 1982): 834-836.

Needham, Joseph. "The Epic of Gunpowder and Firearms." *Chemtech* 13, No. 7 (July 1983): 392-396.

Schwartz, A. Truman, and George B. Kauffman. "Experiments in Alchemy, Part II: Medieval Discoveries and 'Transmutations'." *Journal of Chemical Education* 53, No. 4 (April 1976): 235-239.

Simpson, Christopher. "Military Spin-Offs: By-Products of War." *Science Digest* 90, No. 12 (December 1982): 10-12.

Smith, William B. "Chemistry and the Holocaust." *Journal of Chemical Education* 59, No. 10 (October 1982): 836-838.

CHEMISTRY AND NUTRITION

What We Eat and How It Is Grown and Preserved

Thomas H. Donnelly

Thomas H. Donnelly, Visiting Professor of Chemistry at Loyola University of Chicago, is the author of about 30 technical papers and reviews and the holder of three U.S. patents. Most of his career has been spent in food research and development for Swift and Company, where he served as General Manager, Scientific Services, Research and Development Center, was involved in the development of the "Food For Life" exhibit at Chicago's Museum of Science and Industry and managed basic scientific research for the company. He has served as reviewer for several journals and was on the Advisory Board of the ACS Advances in Chemistry series in 1971. He has been active as an officer in the Chicago Section and the Division of Agricultural and Food Chemistry of the American Chemical Society.

INTRODUCTION

No one should be surprised that chemistry has a substantial impact on our lives in the area of foods and nutrition. After all, we are composed of a variety of vital molecules which must ultimately come from the food we eat, the water we drink, and the air we breathe. From these we must also get our energy. Chemistry contributes to this process by helping to grow our food, by getting it to us, and by helping us understand what we need and how it is used.

CHEMISTRY IN PLANT GROWTH

The production of food begins in the cells of plants and other vegetable organisms. Food provides us with both energy and substance. Thus some foods are more important as energy sources, some as material sources, and most as both. The first step in food production is converting solar energy into food energy. This is done by the process of photosynthesis in which carbon dioxide is taken from the atmosphere and combined with water to produce simple sugars, such as glucose, which are the building blocks of starch, cellulose, and other carbohydrates and which can also be used to provide energy to do work or to synthesize other substances needed by an organism.

In addition to carbon dioxide and water, plants utilize nitrogen, obtained from the atmosphere and usually converted to soluble nitrogen compounds by microorganisms on the roots of such plants as legumes. Nitrogen enables plants to build proteins and nucleic acids, which are chemical compounds intimately associated with living organisms as we know them. Plants also require phosphorus, some of which ultimately becomes part of their nucleic acids.

Potassium is another required plant nutrient which provides an electrolyte necessary for biochemical processes. These three elements, nitrogen, phosphorus, and potassium, are so essential to the nutrition of plants that the numbers by which chemical fertilizers are characterized are measures of them. They can be obtained in adequate amounts from good soil, but such soil is soon depleted of them if cultivation is intensive. Crops such as corn are especially strong consumers of nitrogen, as mentioned above, but legumes, such as beans, clover, etc., can convert the nitrogen of the air into soluble nitrates. Thus, rotation of crops keeps soil from being depleted in nitrogen. Since most of the substance of food-producing plants is not contained in the food produced, the life of soil can be extended by leaving crop residues in the soil.

If crop residues are left, they typically decompose by oxidation, which increases the acidity of the soil. This can be counteracted to some extent by burning residues, which leaves the nutrient minerals in the soil. Soils which have become too acidic for the proper support of plant growth often have their productivity restored by neutralizing the excess acidity with calcium oxide (lime, CaO).

CROP FERTILIZATION

The above simple methods are inadequate to meet the intense food production needs required to feed the more than 4 billion people living on our earth today. Likewise, the practices of early agriculture, including the restoration of nitrogen to soil by using animal manure or decaying animal tissue such as the dead fish used by native Americans, are also inadequate. Even the use of concentrated nitrogen sources such as guano, high-nitrogen bird manure, and nitrate of soda ($NaNO_3$) mined from the Chilean deserts is inadequate. In fact, most of the nitrogen used in fertilizers today is obtained from air by its fixation as ammonia (NH_3) via the Haber process. Ammonia produced by this process can be converted to solid fertilizer components such as urea ($(NH_2)_2CO$) or oxidized to give ammonium nitrate (NH_4NO_3). Today, ammonia itself is taken to the fields as a liquid under pressure. Applied to soil as such, it is the most concentrated source of nitrogen chemically available to plants.

Feeding nitrogen to growing plants in the form of the compounds just described can lead to "burning" of the plants by an osmotic effect of dehydrating plant tissue, especially if all the nitrogen a growing plant will use in an entire growing season is applied at one time. This problem is avoided by the use of "slow-release" nitrogen fertilizers. In fact, the use of decaying animal tissue as a nitrogen source employs this principle. The nitrogen becomes available to plants as it is slowly liberated from the tissue. The same effect is obtained by using animal by-products such as blood meal as fertilizers. Another approach is to make polymers such as those formed by urea and formaldehyde. After application to soil, these polymers are slowly broken down to liberate nitrogen in a form useful for plant nutrition. A current highly acceptable approach is the use of slightly soluble nitrogen compounds (e.g., isobutylidene diurea) which slowly dissolve and are broken down to make nitrogen available to plants.

The demands of plants for phosphorus and potassium in today's high-intensity food production make it frequently necessary to supplement these elements for plant nutrition. In a less intensive type of cultivation, it was possible to obtain phosphorus from the residues of animal bones as "bone phosphate of lime." Today, this need is typically met by mining insoluble calcium phosphate ($Ca_3(PO_4)_2$) deposits and converting them into a soluble form available to plants by treatment with sulfuric acid (H_2SO_4) to produce "calcium superphosphate" (a mixture of calcium dihydrogenphosphate, $Ca(H_2PO_4)_2$, and calcium sulfate, ($CaSO_4$).

The classical source of potassium is the ashes produced by the burning of wood and other vegetation. This is the derivation of the name "potash" (pot ash). The demand for this plant nutrient in high-intensity agriculture also cannot be met by whatever amount of potassium might chance to be in the soil. Again, it is most often met by mining deposits of potassium salts, such as potassium chloride (KCl).

Many other elements, such as sulfur, calcium, magnesium, manganese, zinc, and boron, have been identified as necessary for good nutrition of plants in small-to-trace amounts. These are often made available as small additions to fertilizer preparations, with special formulations used for particular crops when the chemical needs of such crops have been established.

The science of chemistry makes further contributions to food production and distribution by providing the means for minimizing losses in these processes. In the fields where food is produced, herbicides are used to control the growth of plants, such as weeds, which compete with food-producing plants for space, light, and nutrients. In addition, pesticides are used to protect food for our use from consumption by insects and other pests. A modern technology for the control of insects involves the identification, synthesis, and utilization of pheromones, the chemical substances by which insects communicate with each other. The sex attractant pheromones are frequently used to lure insects into a trap where they are destroyed.

In all this, we see that chemistry has had a significant role in identifying substances needed for plant nutrition, in understanding their importance, and in devising ways for supplying these nutrients to growing plants.

PRODUCTION OF FOODS OF ANIMAL ORIGIN

It has been mentioned that all food production begins in the cells of plants. However, some of the foods that we consume do not come directly from plants but from animals which have consumed all or some of those plants. Production of these foods of animal origin, which include milk, eggs, cheese, and butter as well as meats, requires a chemical understanding of animal nutrition. Animal life is characterized in a large measure by its association with protein molecules. Proteins hold animals together, provide them with mobility, catalyze their cellular chemical reactions, and transport oxygen and carbon dioxide, thus performing a myriad of the life functions of animals. These proteins are composed of amino acids. Many animals are not able to synthesize some of the amino acids that they require from simpler compounds in their diets rapidly enough to meet the demands of their own metabolism. These animals must therefore obtain an adequate supply of these "essential amino acids" from their diets. For some animals, this means they must consume the essential amino acids themselves, most likely in the form of proteins. However, animals such as ruminants (cud-chewing quadrupeds) have what may be thought of as more than one stomach, and their various stomachs contain rumen microorganisms capable of converting simple nitrogen compounds such as ammonia and urea into amino acids, thus providing the host animals with an additional source of amino acids essential for their nutrition.

Like plants, animals have other nutritional needs which have been identified, such as a need for various vitamins and trace minerals, along with a need for the chemical energy obtained from their diets. By virtue of identifying and providing all the needed nutrients in their diets in optimum amounts, a situation has developed in which most animals used for food production are nurtured more efficiently than many people.

Chemistry makes yet another significant contribution to the production of foods of animal origin by providing the basis for veterinary medicines and drugs used in the control of animal diseases. Growth promoters of various types have been used in the production of foods of both vegetable and animal origin.

Another significant contribution of chemistry to the system of food distribution is its provision of insecticides, rodenticides, and so forth, by which produced food is saved for human consumption.

HUMAN NUTRITION

Since people are animal in nature, it should come as no surprise that our diets too should be based on the sound nutritional principles of identifying needed nutrients and providing them in the diet. Our diets must provide us with biochemical fuel, building blocks, and substances used in life processes.

The biochemical fuel providing our energy is most often obtained from carbohydrates such as starch and sugar. Fats are richer in energy than carbohydrates and are used by the body to store energy, although some energy is stored as the carbohydrate, glycogen. Proteins are not typically used to provide energy, although this may happen in a starvation situation.

Proteins are the most important of the nutrients serving as building blocks. Just as in the case of animals, there are several essential amino acids which our bodies cannot synthesize from simpler compounds rapidly enough to meet our metabolic demands. For adult humans, these are typically the eight amino acids: isoleucine, leucine, lysine, methionine, phenylalanine, threonine, tryptophane, and valine. In some situations, such as for infants, arginine and histidine are also essential. Of all these, methionine is of great interest because there is frequently a high requirement for it. It is often the single amino acid most limiting a given protein as a source of the essential amino acids. This is somewhat compensated for by the fact that methionine synthesized in the laboratory can be used to satisfy our requirement for this amino acid. This is not true for most amino acids, since they can typically be used only if their molecules are in the same geometric arrangement as that in which they occur naturally. The geometry of the molecules of most of the amino acids is such that their chemical synthesis usually gives rise to equal amounts of molecules of the naturally occurring form and molecules which are the mirror images of this form.

Another class of nutrients which is most important as building blocks are the fat-related lipids which go to make up cell membranes, as well as some of the minerals, such as calcium and phosphorus, of which bones are formed.

Yet another type of essential nutrients are those which provide substances involved in the biochemical processes by which we live. In a way, all nutrients fit into this class, but we are here particularly concerned with those which do not fit into any other class. Vitamins are good examples of these chemical compounds. This name is a contraction of the words "vital amines" which refers to the B-complex vitamins such as niacin, thiamine, riboflavin, and pyridoxine. These compounds typically serve in conjunction with enzymes to help the biochemical reactions of life to proceed.

Vitamin C (ascorbic acid) is a chemical compound for which many attributes have been claimed. A major role which it plays is in the formation of the connecting tissue protein, collagen. This became known after the development of early methods of food preservation made long ocean voyages possible. The collagen disease, scurvy, resulted from inadequate amounts of vitamin C in the diet. This was forestalled by consumption of citrus fruits, such as limes, from which the name "limeys" for British sailors was derived.

Minerals constitute another class of essential nutrients which must be obtained from the food we eat. While our needs for calcium, phosphorus, iron, and iodine have been known for some time, we are only just beginning to learn of our needs for traces of inorganic elements such as zinc, copper, cobalt, and selenium. While elements such as calcium and phosphorus (major constituents of bones and teeth), iron (which helps carry oxygen to our muscles and carbon dioxide away from them) and iodine (essential to the hormone, thyroxine) are easily seen to be essential, even such a common substance as sodium chloride (common table salt, NaCl) is essential to provide the proper ionic environment in body fluids. While our diets seldom contain too little sodium chloride, it does need to be restored to those who lose much of it by heavy perspiration, for example, and persons whose activities result in excessive perspiration often compensate for this loss by taking salt tablets with water.

DISTRIBUTION OF FOOD

In the preceding paragraphs, we have seen how chemistry has helped us to understand the nutrients that we must obtain from our foods, to maximize the production of these nutrients in our foods, and to provide a basis for supplementation if appropriate. But we must remember that food has to be consumed to do any of these things. And before it can be consumed, it must be brought to the consumer. In the agrarian era of our ancestors, it was possible, indeed necessary, for a family to provide all its own food. This was almost a full-time occupation for people in those times. Today, all the food we need is produced by a very small percentage of people. Thus it it essential that we have systems to get this food from its producers to consumers. Because what is food for us is also food for the ubiquitous microorganisms around us, this frequently means there is a need for methods of food preservation.

FOOD PRESERVATION

Until the development of refrigeration and its widespread use earlier in this century, a process in which chemistry made its contribution by providing refrigerants, the preservation of foods often required the use of chemical additives or processing.

Drying is an early, even prehistoric, method of food preservation. The basis for this method is the fact that microorganisms require water to support their growth, and this water is simply not available in dried foods.

An extension of preservation by drying is food preservation by salting. Indeed, any substance which dissolves in water lowers its availability to microorganisms. For this reason, foods preserved by drying are typically not anhydrous (completely water-free) but contain very concentrated solutions of substances with small molecules which are naturally part of the food. This effect is enhanced by adding ionic substances such as salt, whereby the effect is accentuated not just because of the small molecules but also because these small molecules split into ions (charged atoms or groups of atoms) to make the limited amount of water present even less available to support the growth of microorganisms.

At some time in the early history of food preservation by salting, someone used a salt which was or which contained sodium nitrate ($NaNO_3$). By the action of microorganisms present, some of the nitrate (NO_3^-) was reduced to nitrite (NO_2^-). This became a source for nitric oxide (NO), which reacted with the pigments of blood and meat to form a pink pigment which would no longer oxidize easily, thus removing a means by which oxygen could be transported through the salted meat. In this way the art of meat curing evolved. Only in recent years has it been found that the direct use of small amounts of sodium nitrite ($NaNO_2$) gives all the benefits and few of the liabilities of the massive amounts of sodium nitrate which had been used historically.

Preservation by salting does not require salt but only substances with small molecules that are very soluble in water. Sugar molecules are responsible for the fact that candies and pastries do not spoil easily. Glycerin has molecules which are somewhat like those of sugar. It is used to advantage, along with its close relative, propylene glycol, to provide stable food products such as semimoist pet foods.

Another method of preservation which is satisfactory for some foods is pickling. This relies on the fact that many spoilage microorganisms do not grow well and may even die in an acidic environment. The prototypical example is the preservation, albeit in a modified state, of cucumbers and other vegetables as "pickles." Unfortunately for us, however, some molds and yeasts do grow under acid conditions, so that mere pickling may not be a satisfactory method of food preservation. In these cases, it may be supplemented by antimycotic agents to retard the growth of molds, for example. Among such agents are the propionates, benzoates, and sorbates. To those who object to these as "chemicals in our foods," we must point out that such agents are typically used at very low levels, that they are far less harmful than the toxins produced by molds, and that Swiss cheese contains about 1% calcium propionate produced naturally in the cheese-making process.

Many foods which are low in available water and are thus stable against microbiological spoilage also contain significant amounts of fat. (Such foods are sometimes classed as "junk" foods. It is our contention that there is no "junk" food, only "junky" ways of using food. As we have already seen, every component of such foods, i.e., carbohydrates, fat, and salt, may be an essential nutrient.) Foods which contain fats are especially vulnerable to spoilage by oxidation. This can be minimized or prevented by adding antioxidants, of which vitamin E, or α-tocopherol, is an example. Vitamin E, however, is much less effective (and much more expensive) than the commonly used antioxidants such as butylated hydroxyanisole (BHA) and butylated hydroxytoluene (BHT). As is the case with antimycotic chemical food additives, chemical antioxidants are probably less harmful than the products, such as the known carcinogen malonaldehyde, produced from the essential fatty acids by oxidation in their absence, and the rancidity-producing fatty acids with small molecules.

PRESERVATION BY PACKAGING

Foods can be preserved for future consumption if they can be put into a package which cannot be penetrated by microorganisms and if the microorganisms already in the food are then destroyed by processing the package after it is sealed. Any such process is now known as appertization, in honor of Nicolas Appert (1750-1841), the inventor of the first such process which involved the sealing of foods into glass jars and subsequent cooking. Until Appert's invention during the Napoleonic wars, the only commonly used methods of food preservation were the methods of limiting water activity already described. The commonest form of appertization is canning, in which the most usual method of destroying the microorganisms is cooking. The contribution of chemistry to this process is in optimizing the properties of the packages. A metal can, for example, is now lined with a polymer coating which prevents food components such as water and naturally-occurring acids from attacking and dissolving the metal. This greatly diminishes the amount of metal needed to make a safe can.

Another application of chemistry to canning is the use of a sacrificial anode, which is attacked by food acids instead of the metal from which the can is constructed. In this way, the can remains intact, and any damage to the canned product by rusting of the can is confined to the area around the anode so that any damaged product may easily be removed and discarded.

A recent advance in appertization is the development of plastic and metal pouches, suitable for packaging foods, which can then be heated to destroy their microorganisms. This results in lighter packages which consume fewer nonrenewable resources and which cost less to ship. Such packages usually are combinations of metal and polymer films designed to provide strength, flexibility, and impenetrability by light and air. Developing suitable sealants for these packages has required the special combination of desirable adhesive properties with edibility (or nontoxicity) of formula components. A similar problem exists in the sealing of cans, which now use a gasket or adhesive instead of solder, which frequently contained objectionable heavy metals such as lead, cadmium, etc.

A modern form of appertization called "radappertization" employs ionizing radiation to destroy the food's ambient microorganisms. Chemistry makes its contribution to radappertization not only by providing packages but also by providing the basis for understanding and controlling other changes which may take place in the food product during the process.

Packaging materials used to extend the shelf lives and to display semiperishable foods represent another contribution of chemistry to food preservation. Products such as processed meats, for example, are frequently wrapped in a package which will hold a vacuum and through which oxygen cannot permeate. This minimizes oxidation processes such as those mentioned previously and thus protects against the development of rancidity and discoloration.

APPETIZING FOODS

As we have seen, chemistry provides us with a basis for understanding the nutritional requirements that our foods must supply as well as for understanding how to optimize the production and distribution of foods. Thus wholesome, nutritious food can be brought to the table at relatively low cost, in great part because of the contributions of chemistry throughout the food chain. But wholesome, nutritious food is of no value unless it is eaten. Many foods are tasteless, colorless powders which have little appeal even, in some instances, to the starving. An important role of food chemistry and technology is making such foods not just palatable but appealing. This involves manipulation of the sensory properties of foods. Among these are aroma, flavor, color, and texture. Of these, color has long been used as an indication of wholesomeness. In addition, the stabilization of color frequently blocks an oxygen transport mechanism which could otherwise contribute to the development of rancidity, as previously mentioned.

Flavor and aroma represent areas in which food chemistry is called on to make additional contributions toward increasing the appetizing properties of foods. Flavor *per se* has been defined in terms of four basic tastes — sour, salty, bitter, and sweet. Of these, the chemical essence of sourness is acidity since titratable acidity appears to be directly related to perceptible tartness. Saltiness appears to be directly related to the concentration of dissolved ions appearing in the mouth, although only sodium chloride (NaCl) is recognized as having a truly "salty" taste, since most other salts have substantial bitterness in their flavor profiles.

While the chemical basis of bitterness is not generally understood, some progress has been made in understanding the bitterness of protein hydrolysates. This has been shown to be related to the water solubility of the constituent amino acids, with the most insoluble amino acids being responsible for the most bitterness. This does not seem to provide the basis for a generalization since many very soluble substances such as calcium chloride ($CaCl_2$) are quite bitter.

Likewise, a general explanation of the chemical requirements for sweetness has not come forth in spite of many attempts, a recent one being the consideration of molecular arrangments in sweet substances as they relate to the shape of the receptor sites on the tongue. It is well known that many carbohydrates, especially the simple sugars, are sweet, but some, such as fructose, are sweeter than others such as common table sugar or sucrose, which is sweeter than glucose. Some proteins, such as monellin, are sweet. Some of the nonbitter amino acids, such as glycine, are sweet. Glycerin, a substance used to control the availability of water in foods, is sweet. Since a common basis for sweetness has not yet been discovered, it is not surprising that most of the synthetic sweeteners, such as saccharin, cyclamate, and aspartame, were found to be sweet by fortuitous, accidental discovery.

It is obvious that there are more taste sensations than can be accounted for as simply sweet, sour, salty, and bitter, or combinations thereof. These more sophisticated flavor notes appear to arise from a combination of the sensory responses of taste and smell. Such flavors are often obtained by using substances such as spices and flavor extracts. There are about 1500 of these flavoring substances generally recognized as safe (GRAS) by the U.S. Food and Drug Administration (FDA). These substances may be simple, pure chemicals such as vanillin, recognizable as the primary constituent of vanilla flavor, or they may be very complex mixtures of substances such as the oleoresins of spice. Whatever their chemical nature, these substances are blended by flavorists to give to foods appetizing tastes which may not be native to an otherwise desirable and highly nutritious food.

A concern regarding the long-term effects of substances used as food additives for preservation of color and flavor has been raised in spite of the fact that life expectancy has increased throughout the recent decades in which the use of such substances has become more apparent. (Lung cancer is the only clearly identifiable cause of death whose rate of incidence is increasing.) An innovative answer to such concern involves the coupling of chemical groups with the desired functionality to substances which cannot be absorbed into the body. This approach, which could be especially effective with colors and preservatives, has had some success with flavors as well.

IMPROVING TEXTURE FOR APPETIZING FOODS

Texture is another area in which nutritious foods sometimes need improvement to increase their acceptability. A good example is found with the vegetable protein products, especially those now being used as meat analogues. The first texturized vegetable protein meat analogues were made from protein fibers prepared by extruding a slightly alkaline slurry of vegetable protein into a bath in which the slurry proteins had minimum solubility. The slurry, or "spinning dope," often included egg white protein as a binder. The fibers produced were assembled to give a structure resembling a muscle. This process gave a product with a texture very much like poultry white meat.

Vegetable protein products currently used as ground beef extenders, etc., are texturized by processes similar to those used in the manufacture of dry cereals. In this case, high protein vegetable meals are used. The carbohydrate fraction of the meal is responsible for the structural network which gives rise to the texture. Such products can have a high degree of consumer acceptance, as illustrated by the imitation bacon bits widely used at salad bars, etc., instead of real, crumbled bacon. Likewise, the ground beef extenders certainly provide nutrition as good as and possibly better than the meat they replace.

Chemistry is also applied to control texture in other food products. An example is ice cream, which can acquire a gritty texture, especially if it is thawed and refrozen. Such grittiness results from the growth of both ice and milk sugar (lactose) crystals. This is minimized by the so-called protective colloid action of water-soluble substances having very large molecules, a class which includes gelatin, algin, and carageenan.

Confections represent another type of food in which texture is one of the sole, major variables, since most confections are sweet and as such represent a reward rather than a dietary essential providing food energy. Texture of confections can vary from the hard, glassy consistency of rock candy, to the soft stickiness of some marshmallow and nougat preparations. Texture is regulated mostly by moisture content along with the relative amounts of the various carbohydrate sweeteners of which it is composed. These sweeteners are table sugar (sucrose), which would be almost the only constituent of rock candy, corn syrup, honey, invert sugar, and today, high-fructose corn syrup. Of these, sucrose is the same chemical substance whether it is obtained from sugar cane or sugar beets. It is obtained from the juices of these plants by drying and crystallizing, and it represents one of the purest chemical substances available on a large scale.

Invert sugar is obtained from sucrose by splitting its molecule in half and obtaining a mixture of equal parts of glucose and fructose. Honey is the natural equivalent of invert sugar. High-fructose corn syrup is nearly equivalent to invert sugar, but it contains some molecules larger than sucrose. It is a product of modern technology produced by using the same biochemical transformations that our bodies use when we get energy from the metabolism of glucose. Corn syrup itself is obtained by splitting the large molecule of corn starch with water, obtaining some glucose, some maltose, which consists of two glucose units, and some other substances containing three or more glucose units.

Some of the effects of these different sweeteners in confections are well illustrated by marshmallows. The typical roasting marshmallow has a relatively soft, uniform, somewhat sticky texture. It is formulated not to dry out by having increased amounts of sweeteners with relatively small molecules, such as invert sugar, and not to have graininess by using minimal amounts of sucrose. In the marshmallow confection known as a "circus peanut," a drier texture with small grains, or crystals, of sucrose is sought. This is obtained by using larger amounts of sucrose and less of invert sugar, for example.

The texture of jellies is also varied by varying the chemistry of the components. These preparations are typically solutions of sweeteners, etc., in water immobilized by a network of large molecules throughout. These large molecules interact with each other through some of their hydrogen atoms in a process known as hydrogen bonding. Increasing such bonding in a given volume of a jelly increases its stiffness. An almost ideal situation exists in jellies made with gelatin, such as desserts. Here the jelly melts at the temperature of the mouth and does not have to be chewed apart or diluted to soften as is the case with jellies based on pectin or agar.

Chemistry is also the basis for the manipulation of the properties of substances such as oils, lard, and shortenings. With lard, for example, which separates into fat and oil portions near room temperature, separation can be eliminated by producing a substance with what is called a different crystal habit. The molecules of which these substances are composed are E-shaped, with three fatty acids being hung onto a glycerin backbone. Rearranging the fatty acids changes the crystal habit, and a suitable rearrangement can produce a lard from which oil does not separate.

Chemistry also gives these products the ability to make an oil into a shortening. This requires adding hydrogen atoms to the oil molecules. This process, called hydrogenation, along with the possible rearranging mentioned above, makes it possible to make shortenings from almost any fat or oil source.

One final area in which chemistry is employed to improve the texture of foods is in the tenderness of fresh meats. This attribute, after wholesomeness and flavor, is the one most desired by consumers. By the application of chemistry, almost any fresh meat can be tenderized. In fact, meats which used to be edible only as pot roasts can now be cooked and eaten as steaks after broiling. This is all accomplished by developing practical procedures for the isolation and use of natural meat tenderizers, such as those which occur in the meat itself or in papaya or pineapple.

CONCLUSION

From the above, we see that chemistry affects food and nutrition from production to consumption. Our foods are characterized by their chemical nature, as are our needs for them. Our required nutrients include the amino acids, chemically joined together in proteins, the essential fatty acids from oils, the carbohydrates which provide energy, and the vitamins and minerals. In addition, chemicals are used to improve the efficiency of food production, distribution, and consumption, and additives are used to make food more appetizing. All of these contributions of chemistry assure that high quality of food is available to the greatest number of people at the lowest cost.

SUGGESTED READINGS

BOOKS

Benarde, Melvin A. *The Chemicals We Eat*. New York: McGraw Hill, 1971.

Darby, William J., Paul Ghalioungui, and Louis Grilvetti. *Food: The Gift of Osiris*. New York: Academic Press, 1977.

Lee, Frank A. *Basic Food Chemistry*. Westport: AVI, 1975.

Meyer, Lillian H. *Food Chemistry*. Westport: AVI, 1978.

National Academy of Sciences. *Recommended Dietary Allowances*. 9th ed. Washington, DC: NAS, 1980.

Potter, Norman N. *Food Science,* 3rd ed. Westport: AVI, 1978.

U. S. Senate, Select Committee on Nutrition and Human Needs. *Dietary Goals for the United States.* 2nd ed. Washington, DC: Government Printing Office, 1977.

Whitney, Eleanor N., and Corinne B. Cataldo. *Understanding Normal and Clinical Nutrition*. St. Paul, Minn.: West, 1983.

Wenck, Dorothy A., Martin Baren, and Sat P. Dewan. *Nutrition*. Reston, Va.: Reston, 1980.

ARTICLES

Damon, G. Edward. "Primer on Food Additives." *FDA Consumer* 7, No. 4 (May 1973): 10-16.

Foster, Edwin M. "Is There a Food Safety Crisis?" *Nutrition Today* 17, No. 6 (Nov./Dec. 1982): 6-14.

Jukes, Thomas H. "Organic Food." *CRC Critical Reviews in Food Science and Nutrition* 9 (1977): 395-418.

CHEMISTRY AND ENERGY

Solving Short-Term and Long-Term Needs*

Kenneth E. Cox

The late Dr. Kenneth Edward Cox, a noted chemical engineer in the field of hydrogen energy, was born in Tianjin (Tientsin), China on May 5, 1936. He received his B.Sc. from Imperial College, University of London (1956), his M.A.Sc. from the University of British Columbia (1959), and his Ph.D. from Montana State University (1962). He was a research engineer with the Dow Chemical Company, Walnut Creek, Calif. (1962-65), Professor of Chemical Engineering at the University of New Mexico (1965-77), and Project Manager of the program for the development and evaluation of thermochemical processes for hydrogen production, Los Alamos Scientific Laboratory (1977-80). As the author of more than 50 articles, he earned international recognition for his work and received frequent invitations from around the world to lecture on the production, utilization, and economic aspects of hydrogen. He was Editor-in-Chief of "Hydrogen, Its Technology & Implications" (1976-79), a member of the Thermochemistry of Hydrogen Review Panel, U.S. Depart- of Energy (1978-80), and a U.S. representative at the International Energy Agency Technical Workshop on Thermochemical Processes held at Ispra, Italy (1978) and Los Alamos (1979). A week before his premature death in a one-vehicle accident on Pajarito Road near Los Alamos on Sept. 27, 1980, he had returned from presenting a lecture series on hydrogen energy at the Jiaotong University, Xian, China.

---

* This work (LA-UR-79-118) was performed under the auspices of the U.S. Department of Energy. Although the editors have added more recent books and articles to the late Dr. Cox's list of "Suggested Readings," they have not made any substantial changes in his text.

Energy is defined as the capacity to do work. Cheap, abundant energy, in the form of work and heat, has historically been taken for granted by most Americans. In part, this traces back to a persistent pioneer mentality, to the unbounded days of frontier life, and to exploitation without concern for the future.

A review of the historical record of natural resource depletion during the past century will set our energy use habits in perspective. In Figure 1 the curve labeled total energy is the sum of the heat supplied by burning wood and, initially, coal. Wood was the prime energy source in the Unites States in the mid-1800s, but its use peaked a century ago and therefore appears as only a small pimple on our chart, where the bottom line is energy equal to 100 million tons of coal.

Prior to 1900, the United States shifted to a coal-burning economy and averaged an annual production of about half a billion tons of coal for many years. Beginning with World War I, oil and, to a lesser extent, gas began to provide energy for the nation. Ease in obtaining gas and oil from the ground by means of pipes and the development of pipeline transportation as well as the immense automobile-generated demand for petroleum products sent oil and gas consumption rocketing upward so that by mid-century coal was displaced as the nation's No. 1 fuel.

If we pause at this point to survey how far we have come in this century, it is clear that except for a setback during the Depression, when national energy consumption followed the path of gross national product, the pattern of energy use is one of steady growth, largely as a result of chemical combustion of fossil fuels. This growth produced certain adverse effects which, until quite recently, were tolerated by the public, presumably as a necessary evil accompanying the industrialization of America. From the swift ascent of the oil and gas curve in Figure 1, it is easy to realize that sooner or later the combustion of so much fuel would have environmental effects. Nevertheless, few scientists or technicians foresaw these adverse effects. The burning of chemical fuels, whether as solids in steam boilers to generate electricity or as liquids in internal combustion engines to provide mobility, has emerged as a problem requiring technological solution on a massive scale. We now have an Environmental Protection Agency charged with safeguarding the public health

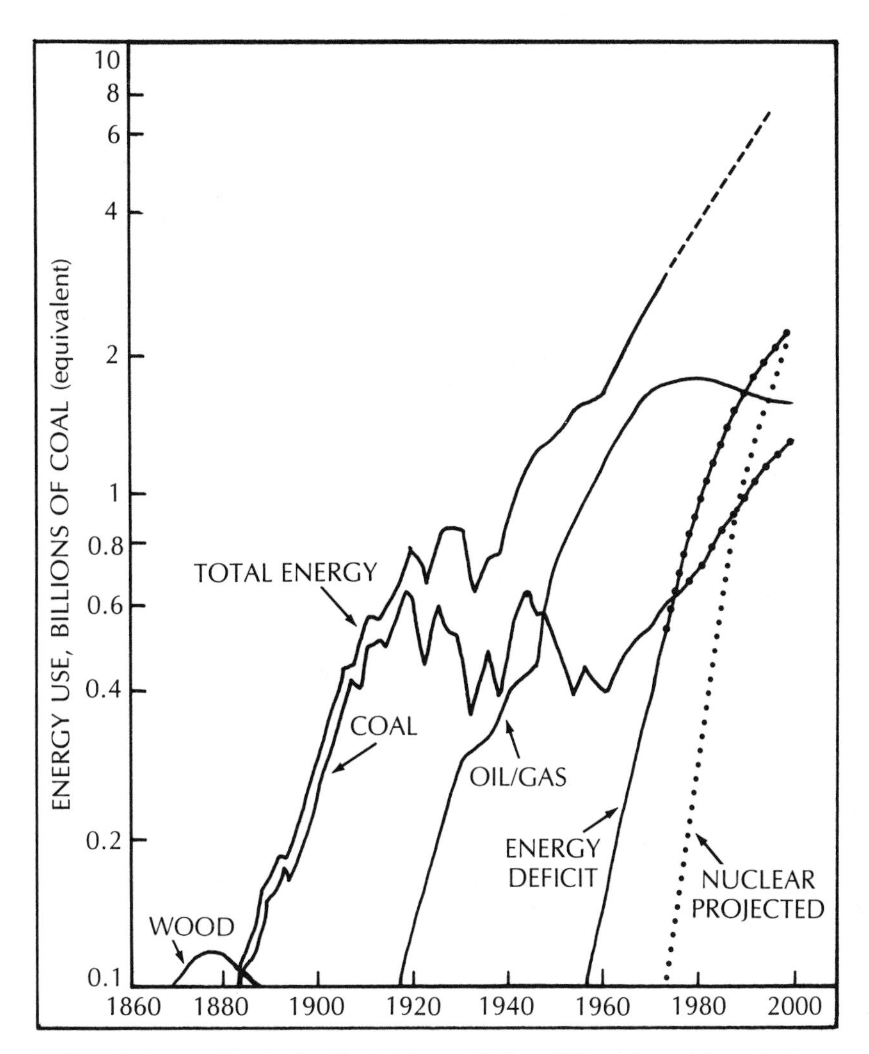

FIGURE 1: Energy Use in the United States 1860 to 2000. Adapted from R. E. Lapp, "The Chemical Century," *Science and Public Affairs: Bulletin of the Atomic Scientists, 29* (7), 8-11 (1973).

from the adverse effects of environmental hazards, and environmentalists must be given credit for focusing national attention on the hazards of fuel combustion.

If we look to the future and extrapolate the national energy consumption curve to the year 2000, there are obviously many curves that we may plot. It can be argued that the United States does not need to sustain such a high growth rate for the next several decades. Certainly, energy is wasted in our affluent society, and there are a number of ways to conserve. However, these conservation methods may be incremental in their effect, and many are difficult to implement in a democratic society. We must therefore find the means to synthesize fuels and chemicals from renewable energy sources in an efficient manner within the next several decades, while preserving our remaining fossil fuel (coal, oil, and natural gas) resources and simultaneously further reducing pollution and wastes. In a nutshell, these are the challenges that face chemists and chemical engineers today as they look forward to the twenty-first century.

CHEMISTRY AND THE NEAR-TERM ENERGY SUPPLY

The President's National Energy Plan, delivered in 1977, spells out in detail near-term energy policy (until the year 2000). Coal and nuclear energy are destined to play larger roles in the nation's energy future than at present. These energy sources will be used primarily to replace dwindling supplies of natural gas and oil and to aid in reducing our dependence on petroleum imports.

National energy consumption is expected to rise from a total of 75 quadrillion BTU (equivalent to 3 billion tons of coal) in 1977 to 130-170 quads by the year 2000. (One quad $= 10^{15}$ BTU [British thermal units], equivalent to 40 million tons of coal $= 10^{18}$ joules. One BTU is the quantity of heat required to raise the temperature of one pound of water one degree Fahrenheit. One joule is the amount of work done when the point of application of one newton is displaced a distance of one meter in the direction of the force.) However, recent work at the Institute for Energy Analysis at Oak Ridge, Tennessee suggests that standard projections of energy demand are far too high because they do not take into account changes in demographic and economic trends. With a vigorous conservation program the likely range of the nation's energy demand in the year 2000 might be 100-125 quads, with the lower value being more probable.

The projected energy needs imply an increase in coal consumption from a total of 600 million tons in 1975 to at least

1200 million tons per year. Roughly two-thirds of the coal will be used for electrical power generation. The remaining portion will be gasified or liquefied to provide gaseous or liquid fuels for use in heating and transportation.

The implications of this greatly increased coal consumption rate are that times of resource depletion will no longer be measured in centuries but in tens of years. Again, certain estimates suggest exhaustion of our known coal reserves as early as 2030 or no later than 2100. Land degradation, air and water pollution, and unknown climatic consequences due to the increase in carbon dioxide and particulate concentration in the earth's atmosphere are other factors relevant to increased coal consumption.

Nuclear energy is at present responsible for less than 12 percent of the nation's electrical generating capacity. Limitations to the increased use of nuclear power have been sought on several fronts, not the least of which is the opposition to its continued use on moral and ethical grounds by a highly vocal segment of the population both in the USA and abroad. More important, however, is the shortage of uranium ore predicted to occur as early as the 1980s.

Nuclear power policy, addressed by President Jimmy Carter in 1977, contains two major themes in nuclear power development: concern with nonproliferation and emphasis on alternative fuel cycles. These lead to the conclusion that nuclear energy cannot wholly fill the energy gap caused by the relative decline in the supply of conventional fuels such as oil and natural gas.

What are some of the steps that chemists and chemical engineers can take in this transitional period? Obvious ones are deriving ways to substitute gaseous and liquid fuels derived economically from coal for expensive imported oil and liquefied natural gas (LNG). Several paths to these goals are being investigated.

CHEMISTRY'S CONTRIBUTION TO THE NEAR-TERM FUTURE

Alternative Fuels from Coal

It is widely recognized that at present synthetic fuels derived from coal are in general not economically competitive. Processes for synthesizing fuels from coal do work, especially gasification processes to make a low- or medium-heating-value gaseous fuel.

G. A. Mills of the Fossil Energy Division of the Department of Energy has concluded that coal gasification plants are too costly because of expensive equipment, excessive complexity, high operating pressure, low throughputs, and excessive consumption of hydrogen, especially in liquefaction processes. He raises the fundamental question of whether or not this is inherent in the differences in the composition of coal and petroleum. Mills' answer is negative. While coal has less hydrogen and more undesirable sulfur, nitrogen, and oxygen than clean-burning petroleum fuels, chemistry ideally should be able to find process improvements or new processes for making synthetic fuels under milder and more selective conditions and, consequently, at lower prices than conventional processing.

Present-day coal gasification processes show the wide difference between what we want to do and what we know how to do:

IDEAL

(1) $2C + 2H_2O \rightarrow CH_4 + CO_2$

Coal Water Methane Carbon Dioxide

GASIFICATION

(2) $3C + 3H_2O \rightarrow CH_4 + CO_2 + H_2 + CO$

Coal Water Methane Carbon Dioxide Hydrogen Carbon Monoxide

(HIGHLY ENDOTHERMIC OR HEAT-ABSORBING)

SHIFT

(3) $CO + H_2O \rightarrow H_2 + CO_2$

Carbon Monoxide Water Hydrogen Carbon Dioxide

METHANATION

(4) $CO + 3H_2 \rightarrow CH_4 + H_2O$

Carbon Monoxide Hydrogen Methane Water

(HIGHLY EXOTHERMIC OR HEAT-RELEASING)

Equation 1 states that theoretically coal reacts with water to yield equal parts of methane and carbon dioxide. Note that this reaction is balanced not only in terms of material but, more importantly, in terms of heat requirements. Thus it should transform coal to methane with essentially no heat loss.

In an actual gasification process, reactions shown by equations 2 to 4 are employed. Because reaction 1 is slow, the coal must be heated to temperatures above 927° C. (1700.6° F.) in order to gasify it. The methane produced is not stable at these temperatures, and thus a mixture of methane, carbon monoxide, carbon dioxide, and hydrogen is obtained (reaction 2). Energy is needed not only to raise the temperature but also because reaction 2 is endothermic (heat-absorbing). To obtain methane, reactions 3 and 4 are necessary. While methanation (methane production; reaction 4) generates some heat, this heat is usually rejected because its temperature is too low (227°-327° C. or 441°-621° F.) for either process heat or electricity generation.

Thus there is apparently much room for improvement in current coal gasification processes. Research is under way to define catalysts (substances that accelerate the rate of a chemical reaction without being consumed in the process) to speed up reaction 1. Researchers at various laboratories under DOE sponsorship have found that certain alkaline catalysts, notably sodium and especially potassium oxide, speed up the steam-coal reaction greatly. Pilot-plant work is under way at Exxon Laboratories to define yields and to determine ways of recovering and recycling the catalyst.

A second example of a new process involves the manufacture of methanol (wood alcohol, CH_3OH) from coal. As already mentioned, carbon monoxide and hydrogen are produced in all coal gasification processes. These gases can easily be converted to methanol with a suitable catalyst. The use of methanol as a liquid fuel or as a basic chemical intermediate needs no elaboration. There has been, however, a technical breakthrough by chemists at the Mobil Oil Laboratories, with DOE support, in turning methanol directly into gasoline of high-octane rating. The process is simple; methanol is passed over a catalyst and is converted almost quantitatively to gasoline and water. The key to this process lies in directing the selectivity of the catalyst by controlling the size of the catalyst pores. A new class of Molecular Sieve ® catalysts (synthetic zeolites with interstices of specific dimensions) were discovered; these catalysts allowed the passage of molecules as large as gasoline molecules but no larger. The process is closely related to the classical Fischer-Tropsch process used extensively in South Africa to make gasoline from coal. The Mobil process has, however, two notable improvements. It is much more selective in making gasoline, and its product has a higher octane number because of its higher content of aromatic substances (compounds related to benzene), typically, about 40%. Exploratory work is now under way to go directly from the synthesis gas mixture, composed of carbon monoxide and hydrogen, to gasoline. Thus chemists and chemical engineers are seriously engaged in developing new processes to make synthetic fuels from coal economically.

CHEMISTRY'S CONTRIBUTION TO THE LONG-TERM FUTURE

What of the long-term future beyond the year 2000 or 2025? Obviously, alternative energy sources must be discovered and developed for this period. In order to avoid the pollution and land degradation encountered with fossil forms of energy, these alternative sources must be abundant, clean in use or with a low potential for pollution, and relatively easy to use at a reasonable cost. More important than these, however, the major attribute of any future alternative energy source should be its renewable nature. Fossil energy or nonrenewable energy sources can be thought of as "capital." When the capital is spent, one must live on "income" or renewable energy. This is the only long-term hope for the future.

Solar Energy

The most obvious energy source which possesses the above attributes and which is perfectly renewable is, of course, the sun. The sun's radiant power is 1.7×10^{17} watts (17 followed by 16 zeros) at the low power density of about 1 kilowatt per square meter perpendicular to the sun's rays on the earth's surface. Obviously, this solar resource is abundant. The present total U.S. electrical demand could be met by conversion plants operating at 30% efficiency using sunlight collected from an area equal to that of the roofs of all of the buildings in the nation.

Use of solar energy has not been extensive, largely because of the cost of its collection, conversion, and storage. As fuel costs escalate, solar energy is becoming increasingly more cost-effective for a number of heating applications (residential heating, hot water, swimming pools, etc.) in the Sun Belt (the southern and southwestern parts of the United States). No doubt, the use of solar energy will continue to grow for these and for other applications as conventional energy costs continue to rise.

To use solar energy as a major source of energy it should be storable in a concentrated form that can be easily transported. This has been the traditional case with fossil fuels, which represent solar energy that has been accumulating over much longer periods of time, i.e., millions of years. Faster and more efficient methods of converting solar energy to a form of chemical or stored energy are clearly needed.

Proposed Methods of Solar Energy Conversion

Man's use of solar energy is not new. Historically, the sun has been used for drying salt and crops. Wind, a manifestation of solar energy, has been used to move sailing ships and provide mechanical power in windmills. Photosynthesis, a natural process, has provided humanity with vegetation as a source of food and fuel. Rainfall, caused by the solar evaporation of the oceans, is used to irrigate crops and to generate hydroelectric power.

Despite these examples, solar energy has not been used on a massive scale for the generation of energy to supply man's total needs due to three major problems encountered in practice. The first concerns the already mentioned lower power density of solar radiation; the second problem is the intermittency of this source. A third problem of practical significance is that solar-derived terrestrial temperatures are low on a scale normally encountered in conventional power-generating systems. Most uses of solar energy require some form of energy storage as a result of these problems. The cost of solar energy systems is thus largely one of overcoming the allied problems of collection, conversion, and storage.

Hydrogen Production

A system that appears to offer high potential for solar energy conversion and storage involves the decomposition of water into its elements, hydrogen and oxygen. Hydrogen is a substance that can be stored in a number of ways, transported with relative ease, and used either by direct combustion to supply thermal energy or converted to electricity via fuel cells (cells producing electricity directly by reaction of a gas or liquid fuel supplied to one electrode and oxygen or air supplied to the other). Furthermore, since water is the only combustion product, the use of hydrogen would be environmentally acceptable and would help to alleviate the world's growing air pollution problems. The use of hydrogen as an energy medium, transferring energy from abundant, clean, massive sources such as the sun to a basic chemical intermediate used in industry, transportation, and households, has been described as the "solar-hydrogen economy."

Serious efforts are presently being made to assess technically and economically solar and other alternative energy sources. These other sources include geothermal heat, gravitational sources (tides), waste materials, etc.

These alternative sources suffer from the same problems as solar energy, i.e., low power density and intermittency. Hydrogen has therefore been suggested as the universal fuel that acts to concentrate and store energy from all sources prior to use.

In his classic book of 1964, *Direct Use of the Sun's Energy*, the late American physical chemist Farrington Daniels (1889-1972) stated:

> One of the best regions of the world for solar applications is the desert of North Chile. Theoretically this waste area receives annual solar heat greater than all the heat produced in the world by fossil fuel burning. It is clear that "theoretically" this desert could supply all the energy needs of the world. The solar radiation could be used to produce hydrogen by the electrolysis of water and the hydrogen could be stored and transported through pipelines or combined with carbon dioxide to give methanol or other transportable fuels. Though not economical now,

these possibilities deserve research study for use in the future.

Daniels' conclusion is still true today.

Conversion of the collected solar energy to hydrogen may be done in a variety of ways. Electricity may first be generated and hydrogen produced by water electrolysis; alternatively, the direct thermal energy from the sun may be concentrated to give a high-temperature (727°-1227°C. or 1341°-2241° F.) heat source, which is then used to drive a thermochemical cycle for water decomposition to produce hydrogen. Photoelectrolytic methods have also been proposed. These methods and some of their accompanying research will now be described briefly.

Electrolysis of Water

A compound may be separated into its constituent elements by heat (pyrolysis), electrical energy (electrolysis), or radiant energy (photolysis or photoelectrolysis). Some of this energy is stored and can be released later by the recombination reaction. Water is a prime example of a compound that may be employed for energy storage by use of its decomposition reaction into hydrogen and oxygen.

$$\underset{\text{Water}}{H_2O\ (\text{liquid})} \rightarrow \underset{\text{Hydrogen}}{H_2\ (\text{gas})} + \underset{\text{Oxygen}}{\tfrac{1}{2}\, O_2\ (\text{gas})} \qquad (5)$$

In the electrolysis process voltage is applied between two metallic electrodes (conductors through which an electric current enters or leaves an electrolytic cell) separated by an electrolyte, i.e., a conductor of ions but not of electrons (electrically charged atoms or groups of atoms). The electrolysis reaction proceeds by electron transfer between the electrodes and the mobile ions in the electrolyte. Gaseous hydrogen (H_2) appears at the cathode (negative electrode), and gaseous oxygen (O_2) appears at the anode (positive electrode). Conventional electrolysis with an aqueous electrolyte takes place at temperatures below 100° C. (212° F.). Problems exist because of overvoltages and slow reaction rates.

Advanced electrolytic techniques use a solid electrolyte, either a plastic membrane (General Electric's SPE or solid polymer electrolyte) or a ceramic substance such as zirconium oxide (ZrO_2). The advantages of a ceramic electrode are twofold: at high temperatures (727° C. or 1341° F.) the rates of the reactions are great, and because of lower voltage requirements the electrical energy consumed is small.

Advantages of electrolysis include clean separations and hence high purity of the hydrogen and oxygen products; no moving parts in the electrolytic cell, ensuring reliability and trouble-free operation; and the provision of an efficient method for generating hydrogen under pressure. The higher voltage requirement for pressurized hydrogen generation is offset by a gain in the cell's efficiency. The major disadvantage of the electrolytic process is the cost of the electrical power required for operation rather than the cost of the electrolyzer. Commercial electrolyzers are capable of operating at electricity-to-hydrogen efficiencies of 70% to 80%. Increases in the efficiency to a range of 90% to 100% are expected with advanced electrolyzers.

In many ways, water electrolysis is ideally suited for coupling with renewable energy sources. As examples, consider solar photovoltaic (providing an electric current under the influence of sunlight) energy and wind energy. The electrical power derived from these sources is highly variant with respect to time. Electrolyzers can adjust to these conditions almost ideally on a demand basis. Transient behavior need not require the use of a control element. Indeed, electrolyzers are self-controlling and inherently stable when operated with a variable power supply. Work performed at the Jet Propulsion Laboratory, California Institute of Technology, Pasadena and at the University of New Mexico, Albuquerque has verified these results.

These experiments have demonstrated the technical feasibility of generating hydrogen from an intermittent energy source such as solar radiation. The main disadvantage of processes of this type is the high capital cost of photovoltaic cells and the lower capacity factor (0.3 to 0.4) of solar energy systems as compared to conventional plants (capacity factors 0.7 to 0.8). Research is presently being carried out to develop ways to reduce the capital cost of photovoltaic cells. A DOE goal is $0.50 per peak watt by 1985.

Photoelectrolysis

Irradiation of an electrode in an electrochemical cell can induce the flow of current in an external circuit. In fact, conventional electrolysis schemes can be assisted photochemically. This technique, although known for the past hundred years, has not received much attention until recently. Sudden interest has arisen in the process of conversion of solar energy to chemical energy.

Photoelectrochemical effects can be obtained with metal electrodes, but much better results are obtained with semiconductor (a substance whose conductivity is intermediate between that of a metal and that of an insulator) electrodes. In the early 1970s Fujishima and Honda made the first claims that a semiconductor-based photoelectrochemical cell could be used for the decomposition of water. Presently, there are more than twenty independent groups in the world actively pursuing this new field of research.

In the photoelectrolysis scheme for the decomposition of water, light is absorbed in semiconducting electrodes, producing electron-hole pairs which are subsequently separated by the semiconductor-electrolyte junction and injected at the cathode and anode to produce reduction (gain of electrons) and oxidation (loss of electrons) reactions, respectively. Hence, an overall reaction is achieved in two steps: (1) electrons and holes are first created by photoexcitation of semiconducting electrodes, and (2) the electrons and the holes drive chemical reactions in an electrochemical cell. These half-reactions are shown by the following equations:

$$(6)\quad \underset{\text{Water}}{H_2O} - \underset{\text{Electrons}}{2e^-} \rightarrow \underset{\text{Oxygen}}{\tfrac{1}{2}O_2} + \underset{\text{Hydrogen Ion}}{2H^+} \qquad \text{Oxidation}$$

or

$$(7)\quad \underset{\text{Hydroxide Ion}}{2OH^-} - \underset{\text{Electrons}}{2e^-} \rightarrow \underset{\text{Water}}{H_2O} + \underset{\text{Oxygen}}{\tfrac{1}{2}O_2} \qquad \text{Oxidation}$$

and

$$(8)\quad \underset{\text{Hydrogen Ion}}{2H^+} + \underset{\text{Electrons}}{2e^-} \rightarrow \underset{\text{Hydrogen}}{H_2} \qquad \text{Reduction}$$

or

(9) $2H_2O + 2e^- \rightarrow 2OH^- + H_2$ Reduction

Water, Electrons, Hydroxide Ion, Hydrogen

Addition of either equations 6 and 8 or equations 7 and 9, respectively, yields equation 5, which depicts the decomposition of water into its constituent elements, hydrogen and oxygen.

Experimental results to date have not been too promising; efficiencies of up to 20% have been obtained with monochromatic light (light of a single wavelength), but with use of solar energy the efficiency dropped to approximately 1%. Continued studies of new materials, chemically modified surfaces, and electrode reactions may lead to useful systems in the future. Since the cost of preparing the materials may be lower than for photovoltaic material and since the need for solar-produced hydrogen is great, there is ample motivation to search for semiconductor materials that more nearly satisfy the stringent requirements in electronic properties than do the materials presently being used.

Thermochemical Decomposition of Water

Since the early 1970s, considerable interest has been shown worldwide in thermochemical methods to produce hydrogen by the decomposition of water. Thermochemical hydrogen production is a means of decomposing water into its elements, hydrogen and oxygen, by a series of chemical reactions involving intermediate chemical species. These intermediate compounds are recycled internally within the process or "cycle" so that water and thermal energy are the only inputs and hydrogen and oxygen are the only outputs. The cycle must be driven by an external heat source such as a central receiver solar collector. The maximum temperature of the heat source for most proposed cycles is in the range of 727°-1027° C. (1341°-1881° F.), eliminating many lower-temperature heat sources as potential suppliers.

The prime purpose for pursuing thermochemical methods for the production of hydrogen is the promise of a higher conversion efficiency of heat to hydrogen than is obtained by alternative methods such as the electrolysis of water. The electrolysis of water requires a power cycle encompassing the inherent Carnot limitation of heat-to-work transformation. Although no thermochemical cycles are as yet commercialized, there is reason to believe that thermochemical cycles can be devised within the range of current or near-term technology. This belief is based on laboratory studies which have demonstrated all of the individual steps comprising a single cycle. Promising cycles are being investigated at United States laboratories including the Westinghouse Research Laboratories, the General Atomic Company, the Los Alamos Scientific Laboratory, and the Institute of Gas Technology. Research in this field is also being conducted in Western Europe and Japan.

The overall heat-to-hydrogen efficiency is expected to be in the 40% to 50% range for thermochemical technology as compared to 25% to 30% for the electrolysis of water, assuming an efficiency of 33% for electricity generation.

Major technical problems with any cycle include heat transfer, heat recuperation, and chemical species separation. Losses of the intermediate species, even at a very low level (0.001%, based on the hydrogen production rate), may seriously affect the economics of the process, because hydrogen is a low-cost commodity item. Engineering approaches are being taken to solve these difficulties so that thermochemical cycle performance may be demonstrated within 10 to 15 years.

The General Atomic cycle and the Westinghouse cycle are examples of these approaches. The General Atomic cycle is characterized by three basic chemical reactions:

(10) $2H_2O$ (Water) + SO_2 (Sulfur Dioxide) + xI_2 (Iodine) → H_2SO_4 (Sulfuric Acid) + $2HI_X$ (aqueous) 27° C. (81° F.)

(11) H_2SO_4 (Sulfuric Acid) → H_2O (Water) + SO_2 (Sulfur Dioxide) + ½ O_2 (Oxygen) 827° C. (1521° F.)

(12) $2HI_X$ → xI_2 (Iodine) + H_2 (Hydrogen) 227°-327° C. (441°-621° F.)

The cycle was made practical by the discovery that if excess iodine is added to the first reaction, a stratification of the liquid phases occurs, yielding an HI_x solution which is immiscible (incapable of being mixed) with the sulfuric acid solution. The acids are readily separated and thermally decomposed to obtain the hydrogen and oxygen products as well as the iodine and SO_2 for recycling.

The Westinghouse hybrid sulfur cycle is a two-step process based on the electrolysis of sulfur dioxide to produce hydrogen as shown by the following equations:

(13)	$2H_2O$	+	SO_2	→	H_2SO_4	+	H_2	(Electricity) 27° C.
	Water		Sulfur Dioxide		Sulfuric Acid		Hydrogen	(81° F.)

(14)	H_2SO_4	→	H_2O	+	SO_2	+	$\frac{1}{2} O_2$	827°C.
	Sulfuric Acid		Water		Sulfur Dioxide		Oxygen	(1521° F.)

A solar high-temperature central receiver might supply the heat for the thermal decomposition of sulfuric acid at 827° C. (1521° F.). Solar electric power would be used for the electrolytic reaction. Liquid sulfur dioxide would be stored because the high-temperature reaction proceeds only during the day.

Photosynthetic Methods

Photosynthetic methods, other than the production of biomass (living matter such as trees, crops, manure, seaweed, algae, etc.), have been suggested as an alternative route to the production of hydrogen. A brief analysis of the photosynthetic process has indicated that it is energetically capable of producing hydrogen and oxygen from water.

To produce hydrogen from water photosynthetically, a process is needed which is capable of catalyzing the reduction of hydrogen ions to hydrogen gas. Certain bacteria and algae contain enzymes (complex organic substances originating from living cells and capable of catalyzing chemical reactions), such as hydrogenase and nitrogenase, which have the ability to perform this task. Such organisms do, in fact, produce hydrogen under certain conditions by photochemical and dark-reaction (non-photochemical) mechanisms. This hydrogen is not produced from water but from an energy-rich nutrient substrate supplied by a separate photosynthetic process. The two biological mechanisms for decomposing water into hydrogen and oxygen apparently exist, but in separate organisms. Coupling these two mechanisms efficiently may be an answer to decomposing water directly for the production of hydrogen.

Without doubt, photosynthetic processes offer the potential for converting solar energy to hydrogen or other useful forms of chemical energy. In principle, yields of up to 10% are obtainable. The major unknowns of such photosynthetic processes are the costs of the needed enzymes, such as hydrogenase or nitrogenase, and the actual efficiency of solar energy conversion that can be attained.

Fusion Energy

In additon to the solar resource, thermonuclear fusion as a future inexhaustible energy source appears to have the advantage of a relatively high power density compared to such previously mentioned alternative renewable sources as solar, geothermal, etc. There are no apparent geographic or climatological restraints, and the supply of the deuterium and lithium fuels used in fusion reactors appears to offer no significant resource concern. Evaluative studies of the application of fusion energy to synthetic fuel production, in which the fuel under consideration is hydrogen, are presently under way at the Los Alamos Scientific Laboratory and the Brookhaven National Laboratory. Fusion energy represents an attractive possibility for the twenty-first century, when fusion reactors will have demonstrated scientific and economic feasibility.

Fusion is the process by which lighter elements such as the isotopes (forms of the same chemical element having different atomic weights) of hydrogen are joined together to form heavier elements with a simultaneous release of significant quantities of energy. The amount of energy released is manifested by the difference between the masses of the products and the fuel reactants; this excess energy can be calculated from Albert Einstein's classic equation:

$$E = mc^2$$

where E is the energy released in joules, m is the decrease of mass (weight) in kilograms, and c is the speed of light (3×10^8 meters per second). Although there are a number of light elements that are capable of undergoing the fusion process, the greatest emphasis in the fusion fuel cycle focuses on the two heavy isotopes of hydrogen, namely, deuterium (D) and tritium (T), with atomic weights of 2 and 3, respectively. Figure 2 describes the D-T fusion fuel cycle and its reactants and products in both the plasma (ionized matter) and the blanket.

The fusion of deuterium and tritium is a high-temperature, high-pressure process. The high temperature (typically 100,000,000° C. or 180,000,032° F.) is needed to overcome the coulombic barrier (electrostatic repulsion of charged bodies of the same sign) that naturally exists between the two nuclei of the two fuels. The pressure requirements stem from the basic need to achieve sufficient power density so that ignited or near-ignited conditions exist in the fusing fuels. In addition to sufficient temperature and pressure, the fuels must be held together

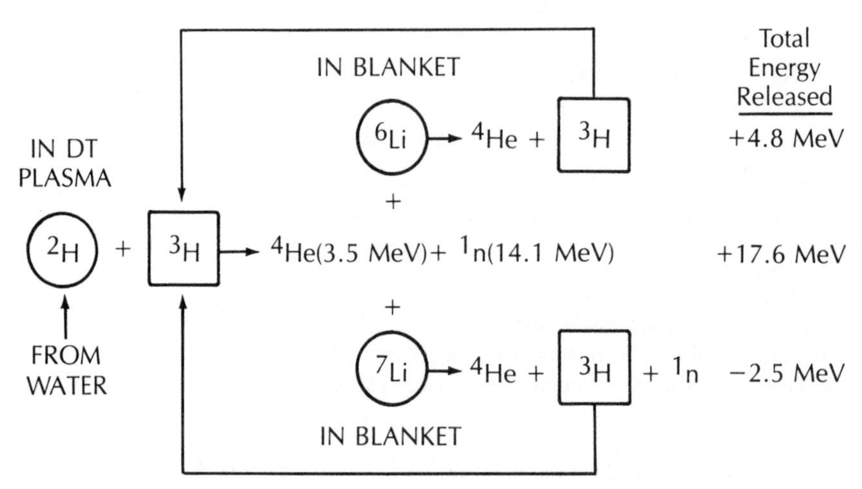

FIGURE 2: The Deuterium-Lithium-Tritium Fusion Fuel Cycle

for a length of time sufficient to allow an adequate number of fusion reactions to take place to satisfy economic requirements.

The fusion program has made remarkable achievements in recent years, and experiments either now under way or being planned are expected to demonstrate the physics of fusion reactor-grade conditions within a few years. Given the success of these experiments — and there is no reason to expect otherwise — a prototype fusion power reactor could be operational before the end of the 1980s.[1]

Fusion research is being carried out following two concepts: principal magnetic fusion and laser fusion. For magnetic confinement, the fuels exist in what is called a plasma, which is the ionized state of matter. Because the fuel nuclei no longer have the associated electrons (ordinary unionized matter consists of protons, neutrons, and electrons), they are affected and confined by imposed magnetic and electric fields. A reactor called a tokamak provides confinement by a toroidal/poloidal magnetic field, which is designed to prevent the fuel particles from reaching the walls. The fuels are heated to reaction temperature by a variety of methods including ohmic collision, radiofrequency injection, and neutral particle beams. Other approaches feature similar confinement and heating methods, the principal differences being in the arrangement of the magnetic fields and the heating rates. In inertial confinement fusion, brief intense pulses of either laser radiation or electron beams or ion beams are focused on small pellets of fusion fuel (for example, pellets consisting of a mixture of deuterium and tritium). The pulse of energy added to the pellet causes it to implode, creating within it the high temperatures and densities required to initiate fusion. The net result is a

[1]On December 25, 1982 scientists at Princeton University announced the world's first successful firing of a new-generation nuclear fusion reactor. The $314 million tokamak fusion test reactor, funded by the Department of Energy, was turned on for the first time on December 24, 1982 after seven years of design and construction efforts. The actual test lasted only five-hundredths of a second, but Dr. Harold P. Furth, Director of the Princeton University Plasma Physics Laboratory, stated that the test was extremely significant despite its short duration. "It's like Columbus finding the New World," he said. "The question is not how big it is, but that they found land."

microexplosion which releases the fusion energy in the form of 14 MeV (million electron volts) neutrons, alpha-particle kinetic energy, pellet debris kinetic energy, and X-rays.

Uses of Fusion Energy

The conventional use of fusion energy in the form of heat would be for the production of electricity. Another attractive application of fusion energy would be the production of synthetic chemical fuels based on hydrogen-from-water decomposition processes such as electrolysis and thermochemical decomposition.

At this time the fusion program has identified high-temperature electrolysis as the hydrogen-producing process that appears the most attractive when coupled with an ultrahot fusion blanket. Thermochemical cycles, including those "hybridized" with an electrolysis step, may also be suitable.

The very high temperatures — above 1500° C. (2732° F.) — offered by the fusion blanket may permit the elimination of steps in thermochemical cycles, thus permitting simpler and presumably more efficient and economical fuel production.

Thus, several avenues of chemical research are being pursued to solve the problems of energy supply for the near-term and the long-term. Chemistry will apparently play a vital role in energy production as well as in energy conversion and storage technology. Hydrogen, the first element of the periodic table, is a leading candidate for the 21st century as an energy supply medium. Making hydrogen from water and using hydrogen to satisfy society's energy needs has been termed the "hydrogen economy." Implementing this concept may well be one of the greatest challenges faced by chemists and chemical engineers as we leave the age of fossil energy and enter the era of renewable energy.

SUGGESTED READINGS

BOOKS

Bockris, John. *Energy, The Solar-Hydrogen Alternative.* New Haven: Yale University Press, 1975.

Bromberg, Joan Lisa. *Fusion: Science, Politics, and the Invention of a New Energy Source.* Cambridge and London: MIT Press, 1982.

Daniels, Farrington. *Direct Use of the Sun's Energy.* New Haven: Yale University Press, 1964.

ARTICLES

Bova, Ben. "Hydrogen Fuel." *The High Road.* Boston: Houghton Mifflin, 1981.

Cole, K. C. "Interview: Robert Bussard." *Omni* 3, No. 4(January 1981): 56-58, 90-92.

Conn, Robert W. "The Engineering of Magnetic Fusion Reactors." *Scientific American* 249, No. 4 (October 1983): 60-71.

Cox, K. E. "Hydrogen from Solar Energy." *Hydrogen, Its Technology and Implications*, Vol. I. Cleveland: CRC Press, 1977, Chap. 5.

Edelhart, Mike. "Fusion Odyssey." *Omni* 3, No. 4(January 1981): 38-42.

Eisenstadt, M. M., and K. E. Cox. "Hydrogen Production from Solar Energy." *Solar Energy* 17(1975): 59-65.

Fickett, A. P., and F. R. Kalhammer. "Water Electrolysis." *Hydrogen, Its Technology and Implications*, Vol. I. Cleveland: CRC Press, 1977.

Fujishima, A., and K. Honda. "Electrochemical Photolysis of Water at a Semiconductor Electrode." *Nature* 238(July 7, 1972): 37-38.

Greenberg, Daniel S. "Fusion Politics." *Omni* 3, No. 4(January 1981): 52-54, 109.

Kidder, Tracy. "Taming a Star." *Science* 82, No.2(March 1982): 54-64.

Landa, Edward R. "The First Nuclear Industry." *Scientific American* 247, No. 5(November 1982): 180-193.

Lapp, Ralph E. "The Chemical Century." *Science and Public Affairs: Bulletin of the Atomic Scientists* 29, No. 7(September 1973): 8-14.

McColm, R. Bruce. "The Business of Fusion." *Omni* 3, No. 4(January 1981): 46-51, 86-88.

McKean, Kevin. "The Mounting Crisis in Nuclear Energy." *Discover* 3, No. 5(May 1982): 20-26.

Rose, Kenneth Jon. "The Hydrogen Man." *Omni* 4, No. 4(January 1982): 50-53, 95.

Solomon, Burt. "Will Solar Sell?" *Science* 82, No. 3(April 1982): 70-76.

Thomsen, Dietrick E. "Economy of Fusion." *Science News* 125, No. 1(1984): 10-11.

CHEMISTRY AND MEDICINE

Modern Advances from Wonder Drugs to Surgical Sutures

Ned D. Heindel

Ned D. Heindel, the H. S. Bunn Professor of Chemistry at Lehigh University, Bethlehem, Pennsylvania, is engaged in research on new diagnostic and therapeutic drugs. He is the author of more than 150 technical articles on synthetic techniques and biological evaluation of potential pharmaceuticals. Dr. Heindel serves as Director of Lehigh's Center for Health Sciences and is also Professor of Nuclear Medicine at Hahnemann Medical University in Philadelphia.

To think of chemistry's contributions to medicine is to think immediately of the miracles of pharmaceutical chemistry. It is probably true that the chemist's most impressive contribution to medicine has been his expansion of the physician's selection of drugs. Even the most chemically-naive layman has heard of "wonder drugs" and has had some exposure to the curative, palliative, and, unfortunately, abusive power of physiologically potent chemicals. A recent American Chemical Society report, *Chemistry in Medicine*, noted that drugs were the third largest item in the U.S. annual health care budget, at $10.6 billion, following only hospital care ($46.6 billion) and physician care ($22.1 billion). Most of the rather substantial increase in human life expectancy in recent years has been attributed to the development of new medicinal agents.

A focus on chemistry's contribution to drug discovery is important but represents only one of the multifaceted ways in which chemistry serves medicine. An admittedly incomplete listing would also include clinical laboratory analyses, diagnostic radiopharmaceuticals, biomedically important plastics, and medical education and research.

DRUG DISCOVERY

An understanding of the biological action of chemicals dates back almost to the dawn of recorded history with references to the uses of the poppy, the cocaine leaf, and the hemp plant

FIGURE 1: Just as man has found numerous therapeutics in plant sources, so has he found hallucinogenic and mind-altering chemicals in natural sources. Specially prepared teas and dried solid portions of cactus buds from *Lophophora williamsii* (common name, peyote) were first used by the Aztecs to induce visions. The peyote buds are still ingested by communicants of the Native American Church as part of their religious sacraments.

found among the earliest folk remedies. Cinchona bark for treatment of malaria dates at least to the sixteenth century. However, the systematic use of pure chemical substances evaluated in a controlled fashion to probe their medical utility is very much a derivative of the efforts of Paul Ehrlich (1854-1915).

The founder of modern chemotherapy, Ehrlich was a late–nineteenth–century physician, microbiologist, immunologist, and chemist. The major killers of his day were the infectious diseases. Louis Pasteur (1822-1895) and Robert Koch (1843-1910) had done the groundwork of identification, cultivation, and classification of microorganisms, and from their work, Ehrlich was well aware of the utility of dyestuffs in bacterial classification. (The classic Gram negative and Gram positive diagnostic dichotomy is based on just such a selective staining technique.) Working with infected live animals, Ehrlich discovered that occasionally a dye would be absorbed into the organism with greater specificity than onto any portion of the animal host. As a result, he suggested his now famous "magic bullet" concept that suitable chemicals might be uncovered with high target specificity which would transport killer moieties to an infected area with minimal damage to noninvolved healthy organs.

At that time, topically applied pastes of heavy metal salts (arsenic, lead, mercury, or antimony) were used for the treatment of the syphilis spirochete. These metal ions, although toxic to organisms cultured outside the body in Petri dishes (the so-called *in vitro* assay) were of marginal efficacy when tried with human sufferers (*in vivo* evaluation). Too many patients experienced the side effect of metal poisoning. Ehrlich reasoned that a spirochete-staining dye bonded chemically to one of these metals might prove a useful pharmaceutical in the treatment of syphilis. After 605 laboratory failures, he proved his theory. Furthermore, his logic and his laboratory method survive today as cornerstones of medicinal chemistry.

Salvarsan (606, arsphenamine) was the 606th chemical synthesized by Ehrlich and his co-workers. It was active *in vivo* against the spirochete and displayed few toxic manifestations in human patients. Salvarsan was used clinically for several decades as the drug of choice for treating syphilis.

FIGURE 2: Paul Ehrlich (1854-1915) chemically combined organic dyestuffs which displayed a staining specificity for selected biological membranes with toxic agents to produce new drugs specifically directed to those target tissues. Dr. Ehrlich is often referred to as the founder of modern chemotherapy.

Photo courtesy of the ACS Center for the History of Chemistry, University of Pennsylvania.

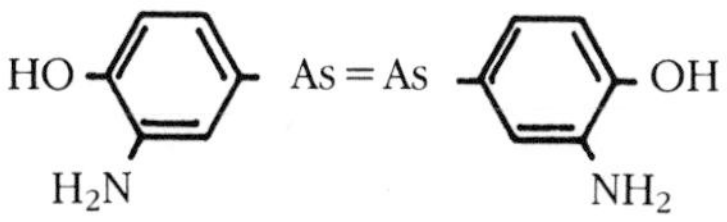

HO — C_6H_3 — As = As — C_6H_3 — OH
H_2N NH_2

ARSPHENAMINE (Salvarsan)

Chemists have long been troubled by the more or less empirical way of seeking new drugs pioneered by Ehrlich. Although some progress is being made in the understanding of the molecular etiology of a disease, i.e., the chemical causes of a disease, and in the design of specific chemicals to inhibit it, the observation by Alfred Burger, foremost communicator of medicinal chemistry today, remains painfully true:

> The *de nova* discovery of drugs is still an unaccountable empirical adventure. One can screen compounds more or less randomly for a given disease entity. Or one can follow therapeutic folklore in natural products research. If one is lucky, one hit in several thousands will guide us to a novel active material.

Although Ehrlich's work was important and widely known (he received the Nobel Prize for Physiology or Medicine in 1908), it was the research of Gerhard Domagk (1895-1964) which laid the groundwork for the modern period of drug discovery. Domagk, Professor of Chemistry at Münster, later Research Director of the I. G. Farbenindustrie Institut in Elberfeld, Germany, and 1939 Nobel laureate in physiology or medicine, headed a team of chemists who, in 1935, synthesized the first widely useful bactericidal agent, the red dye Prontosil. It was soon discovered that Prontosil was reductively cleaved *in vivo* to the true active form, sulfanilamide. Sulfanilamide proved potent against the bacteria responsible for rheumatic fever, tonsilitis, several pneumonias, and puerperal fever. Almost immediately, a fruitful field of medicinal chemistry was launched.

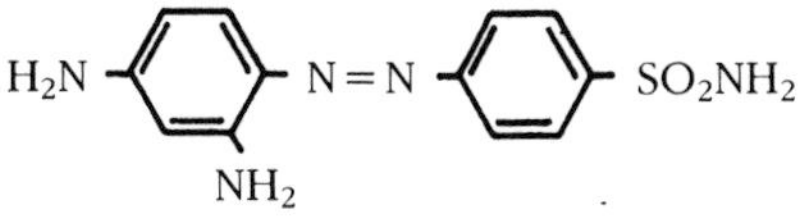

H_2N — C_6H_3 — N = N — C_6H_4 — SO_2NH_2
NH_2

Prontosil

H_2N — C_6H_4 — SO_2NH_2

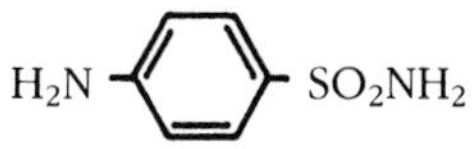

Sulfanilamide

FIGURE 3: Syphilis, a painful scourge of mankind for centuries, was described in several medical texts of the Middle Ages. Physicians' empirical observations that compounds and formulations of heavy metals provided some transitory relief led medicinal chemists of the late 19th and early 20th centuries to focus on drugs containing such metals for systemic treatment of syphilis.

Various experts have estimated that between 5,000 and 10,000 analogues containing the variations on the chemical structure of sulfanilamide have been synthesized and evaluated in the intervening 45 years. Not only have greatly improved ranges of antibacterial potency been discovered, but the fallout from broad-based animal evaluation of these potential sulfa drugs has produced commercial hypoglycemics (the so-called oral insulins), diuretics, antihypertensives, antituberculars, and antileprosy agents. The efforts of the synthetic chemist have truly been well rewarded.

Just as in Ehrlich's day the infectious diseases were the chief causes of morbidity, so today (in the post-sulfa, post-penicillin, post-antibiotic era) the major killers in the United States are cardiovascular disease (979,180 deaths per year in 1975) and cancer (371,740 deaths per year). Here too, the wonders of the chemist's touch have been demonstrated. The search and evaluation techniques of Ehrlich have proven useful in the discovery of cardiovascular drugs intended to lower cholesterol, control cardiac arrhythmias, inhibit clot formation, relieve the pains of angina, reduce hypertension, and alter heart rate. Much has been done but much more remains to be done. Many of today's available drugs show some systemic toxicity, and few are totally effective for all sufferers.

Fewer successes have occurred in the area of cancer chemotherapy. Cancer is not one disease but many, and until recently it remained generally irreversible. Such cures as did exist were made possible only by surgical removal of the unmetastasized primary tumor. Irradiation and chemotherapy appeared to delay death but not to annihilate totally the malignancy itself. Today prolonged remissions are possible in several kinds of cancer because of modern synthetic chemicals.

Cyclophosphamide, a nitrogen mustard alkylating agent whose structure is based on the discovery of antileucocytic properties of the mustard gases prepared for the military in World War II, yields about a 30% cure rate in Burckitt's lymphoma. An antibiotic, actinomycin D, appears able to produce an 80% cure rate or at least very long–term survival rates with Wilms' tumor in children. Many other examples might be cited.

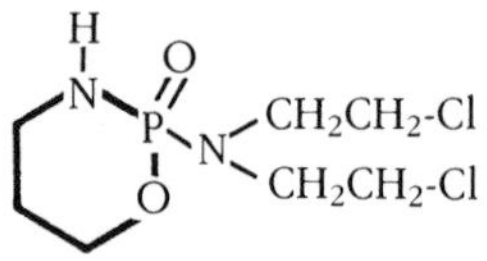

Cyclophosphamide (an anticancer drug)

The medicinal chemist of today seeks anticancer drugs in analogues of structures which have shown previous clinical success. Very often it is found that minor modification in the 3-dimensional shape of a chemical or drug enhances a desired biological effect and suppresses a side-effect (unfortunately, the reverse is often observed too). New potential therapeutics which are close in structure to nature's own building blocks — used by the body in synthesizing protein, fat, bone, and connective tissue — often interfere with the function of the natural biological substances in a beneficial way. A drug which blocks protein synthesis may retard the spread of cancer, while an agent which interferes with fat deposition may inhibit coronary artery disease.

In "Chemistry and Human Behavior: The Next Frontier of Science" (Chapter 12 of this volume) Manfred G. Reinecke has noted that a fascinating interplay exists between chemicals and mental function. Many experts now believe that a chemical imbalance is at the basis of much mental illness. A knowledge of how normal chemicals — the neurotransmitters and hormones — react in the brain is making it possible to prepare synthetic chemicals (drugs) capable of specific behavioral alteration.

A future is dawning in which the medicinal chemist can design drugs from a knowledge of the molecular biology of the target disease. He can take several active leads already in hand and, by use of a computer and a newly developed predictive relationship (the so-called "Hansch equation") between certain physical properties and potency, he can predict a better structure for a previously unsynthesized chemical that should be prepared and tested. The chemist of today does not work alone. He needs the help of biologists, physicians, computer scientists, and pharmacologists. Increasingly he lives in an era of enlightened empiricism.

CLINICAL LABORATORY ANALYSES

The modern physician depends heavily on laboratory results in making the critical decisions of diagnosis and treatment. Through the years, as medicine has become less of an art and more of a science, the hospital clinical laboratory has been at the forefront of that transition. Furthermore, clinical laboratory science has increasingly become applied chemistry. Whether the measurement of biological, chemical, and physical phenom-

FIGURE 4: Biochemists have developed numerous clinical assays which can be automated and performed serially with minimal supervision by sophisticated computerized clinical analyzers. Such equipment significantly extends the range of tests which can be performed by a single clinical technologist and permits high accuracy at a modest cost per assay.

Courtesy of Technicon Instruments Corporation, Tarrytown, N.Y.

ena within the body is done by classic "wet" test tube methods or by complex instruments, it is still chemistry in action.

The combination of sophisticated instruments used in sets has provided powerful new tools for the separation and identification of biologically important molecules present in microsamples of blood, serum, or urine. For example, it is now possible to identify rapidly the particular drug of abuse in cases of overdosing, to monitor blood levels of antidepressants and cardiovascular drugs, and to identify infectious microorganisms by "fingerprinting" their characteristic metabolic products. Earlier methods of pinpointing the particular drug of an overdose or the particular bacterium of an infection were so slow and tedious that treatment decisions were often critically delayed.

While in the past most clinical laboratory measurements depended on color changes produced by the specific interaction between appropriate substances and the key blood enzyme whose concentration was being sought, an increasing number of modern measurements utilize immunospecific reagents. The area of radioimmunoassay (RIA) has virtually mushroomed in the last decade.

Based on the high specificity which results from antibody-antigen complexation, RIA procedures employ a radioactive label on either the antibody or antigen whose concentration is being sought. The accuracy and precision reflected in radioisotopic counting easily permits excellent quantification in the nanogram per milliliter level. Since a nanogram is a billionth of one gram and a milliliter is a thousandth of a liter, such detection levels correspond to accurate measurement of a pinhead amount of an unknown substance dissolved in a quart of fluid. The most popular RIA methods are those used to measure human thyroid hormones, but another important assay is used to monitor blood levels of digoxin, a cardiac drug with an extremely narrow range of therapeutic utility and a toxic level barely twice its therapeutic level. Patients can easily be poisoned by digoxin if its level in their blood is not precisely measured and controlled. Radioimmunoassays have been adopted for numerous other drugs, hormones, and blood proteins.

Twenty-five years ago the physician could call upon his lab for a mere dozen or so blood tests. Today well over 100 different assays are in common use, and there are more than 500 million such lab tests done per year. That number is expected to double in the next five years.

DIAGNOSTIC RADIOPHARMACEUTICALS

Yet another kind of diagnostic assistance to the clinician has come from the hands of the chemist in the form of radioactive pharmaceuticals. Disease-specific molecules are converted into radioactive form by attachment of short-lived, low-exposure radionuclides and are adminstered to the patient by injection, orally, or by inhalation. By virtue of the specificity designed into these radiopharmaceuticals by the chemist, these drugs concentrate, like Ehrlich's "magic bullets," at the sites of malignancy, inflammation, infarction, abscess, or other specific disease. The presence and exact location of such abnormalities can then be noted by external, radiation-detecting, imaging devices which capture a picture of the isotopic distribution on photographic film.

Many diagnoses, which would be impossible or very difficult to achieve by other methods, are possible with radiopharmaceuticals. Tumors of brain, bone, lung, spleen, and liver can be pinpointed rapidly by use of the appropriate radioactive diagnostic. The area of damage to the heart's surface in a recent heart attack can be mapped, giving the physician a measure of the severity of the event. The possibilities of medical diagnoses are impressive and growing rapidly.

Knowing that one of the liver's functions is to clear and excrete large lipophilic (fat-like) dyes, researchers have developed a radioactive version of one such dye (Rose Bengal) for evaluating the extent of liver disease. Similarly, the biological fact that human bone contains phosphate and rapidly absorbs that material from the blood has led to a whole host of radioactive phosphate compounds as differential markers for bone disease. Bone tumors seem to absorb these radiolabeled phosphates with greater rapidity than healthy bone. In a related situation, the adrenal glands have the biochemical role of converting one kind of steroid hormone to another. These glands can absorb cholesterol from the blood and convert it into important biomolecules. This selective uptake has been the basis for the design of a new adrenal-specific radiopharmaceutical (radioiodinated 19-iodocholesterol).

Chemists who work in this growing and important field of radiopharmaceutical science must have an understanding of organic molecules, radioisotopes, physiology, and the unsolved problems of medical diagnosis which the clinician faces daily. The successes have been many, but substantial unsolved problems await the researcher. No suitable specific agents exist for the detection of pancreatic cancer, melanoma of the eye, breast or uterine cancers, or metastatic involvement of the deep-seated lymph nodes of the groin. These are but a few of the new challenges awaiting the scientist. In addition, there is often substantial room for improvement in radiodiagnostics already on the market.

BIOLOGICALLY IMPORTANT POLYMERS

Chemistry has also provided major assistance to the surgeon. In cosmetic, reconstructive, orthopedic, and vascular surgery, materials composed of very large molecules (polymers) not only have made possible operations that were only dreamed of in bygone days but they have also facilitated patient comfort in classical surgical procedures.

Sufferers from degenerative osteoarthritis or aseptic necrosis of joints often faced a lifetime of pain and restricted mobility. Fractures of the hip, knee, or shoulder left patients with permanent reminders of trauma, since incomplete recovery was an all too frequent occurrence. Today, thanks to advances in polymer science, special putties (mixtures of barium sulfate, methyl methacrylate, and suitable initiators) can be employed as physiologically-compatible cements to affix metallic ball joints and sockets to healthy bone. So-called hip joint replacements have been made possible by these fast-setting resins whose physical properties (density, tensile strength, impact resistance, and coefficient of thermal expansion) are equal or superior to that of viable human bone. In most cases the surgeon simply shapes the putty into the desired form to bond the metallic implants to the bone and triggers the solidification (polymerization) of the putty by ultraviolet light. Another application of these methyl methacrylate cements and related monomer putties has been to "glue" fractured bone ends together, thus alleviating the necessity for pins or prolonged immobilization in plaster casts.

Chemistry has even provided a versatile substitute for the traditional plaster cast. More than four million such plaster of Paris casts are applied annually in the United States to repair bone fractures. Complete curing of the plaster usually requires 24 hours. Furthermore, because the plaster can hydrate and disintegrate upon exposure to water, the patient's bathing is restricted, and "windowing" of the cast to permit compressing is a

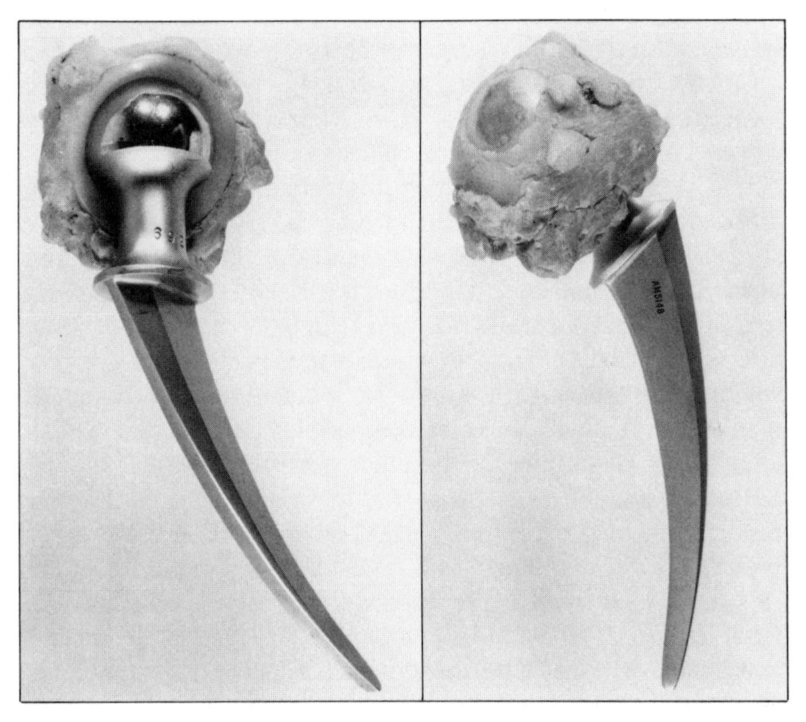

FIGURE 5: A metallic pin of a corrosion-resistant alloy is inserted into the bone of the upper leg and seated into a polymeric cup held in place on the pelvis by a cement of poly(methyl methacrylate). This hip joint replacement, a successful combination of metallurgical and chemical technologies, has been of considerable assistance to sufferers of advanced arthritis or other hip disorders.

virtual impossibility. ("Windowing" refers to a hole left in the cast to apply water soaks to soft tissue wounds near the fracture site.) Now with a new plaster cast these problems have been solved. After placing a sleevelike polypropylene stocking over the fracture site, wrapping the limb with a cushioning web of polypropylene and then with a glass-fiber tape impregnated with photosensitive pre-polymers (a patented mixture of two vinyltoluene-type monomers), the cast is made rigid by photocuring with a lamp emitting light of wavelength in the 3200 to 3900 angstroms (0.0000126 to 0.0000154 in.) range. Compared to the plaster of Paris cast, the polymeric cast is stronger, lighter, faster curing (three minutes), more compatible with water, and less prone to cause itching of the skin.

Chemists have also been at work providing materials for prosthetic and vascular surgery. Silicone rubbers have served for many years as breast implants either for reconstructions following partial mastectomies or for cosmetic purposes. A new woven combination of polytetrafluoroethylene and carbon fibers is proving useful for direct implantation and for anchoring artificial parts. The weblike nature of the product permits the surrounding tissue to grow into it and thus to stabilize the implant, in effect, making it part of the body. This polymer has been employed surgically to correct defects of the bony structure of the jaw, face, and ear, whether these defects are congenital abnormalities or the result of accidental trauma. Sterilizable, resilient, and physiologically compatible, these plastics retain their *in vivo* flexibility for the lifetime of the patient.

These same physical properties have proven useful in the preparation of flexible polymeric blood vessel grafts. Surgical repair of aortic aneurysms and other damaged or diseased arteries and veins is now possible with synthetic vessels. These flexible plastic tubes can be stitched to adjacent healthy arteries and appear to form a permanent knit. Patients with ballooning aneurysms who would have experienced a bleak and shortened future of restricted activities can now be restored to their former vigor.

Surgical sutures which must be left in place often give rise to inflammatory rejection at the site of the surgery. Such so-called "foreign body giant cell reactions," while seldom fatal, can prove uncomfortable and sometimes unsightly. Again, through the wonders of chemical science, biodegradable sutures (made of such materials as polyglycolic acid) have been developed. These sutures are absorbed after sufficient time to allow healing of the incision with minimal tissue reaction. For deep internal surgery, dermatological procedures, and removal of cataracts these soluble sutures are an alternative to the older silk, cotton, or gut filaments.

Plastic films have also been fabricated into extracorporeal membranes to facilitate oxygenation of blood in patients with acute respiratory failure. External membrane oxygenator devices can provide respiratory support like that which is required in prolonged open-heart or open-chest operations. Earlier blood oxygenation machines of the bubble and disk type damaged red blood cells and other blood elements and limited critical surgery to less than six hours. Polymeric membrane oxygenators appear to be usable for much longer periods and can even replace the human lung for sufficient periods to permit its natural recovery from respiratory trauma.

Yet another unique medical application of chemistry is controlling the rate of drug administration to a patient by surgically implanting a large dose of the physiologically active substance encapsulated in a slowly dissolving plastic matrix. Contraceptive drugs, antiepilepsy agents, and narcotic antagonists have been delivered *in vivo* from a variety of patented hydroxylic or amidic polymers. While none of these so-called bioerodible polymers is yet in routine clinical use, a number of different drug-plastic combinations are under study, and a breakthrough appears imminent.

Each polymer for a medical application possesses its own unique combination of physical properties and structure. By understanding the needs of the physician and by being accomplished in the art of structural manipulation of macromolecules, the chemist can tailor a new synthetic to solve the problem at hand.

MEDICAL EDUCATION AND RESEARCH

> I will look upon him who shall have taught me this art even as one of my parents .... I will impart this art by

> precept, by lecture, and by every mode of teaching .... With purity and holiness will I pass my life and practice my art.
>
> Hippocratic Oath

Certainly the physician of ancient Greece had ample reason to refer to his practice as an art. In fact, until recent times, medicine was more of an art than a science. Few today, however, would deny that modern medicine is far more of a science than an art. One of the disciplines most responsible for this transformation in medicine has been chemistry. Physicians today are taught the subjects of oxidative phosphorylation, nucleic acid biosynthesis, the function of the sodium-potassium pump, the molecular origins of immunity, the chemical basis for the actions of pharmaceuticals, the maintenance of physiological *p*H, and a myriad of other chemical topics which provide a scientific rationale for accepted clinical practices.

Through his training in chemistry and biochemistry the physician of today is far more likely than were his predecessors to deduce new therapeutic applications and to make novel contributions to his discipline. The medical student of the 1980s is taught substantial chemistry as it applies to pharmacology, physiology, pathology, and biochemistry. In addition, he has been taught inorganic, organic, and analytical chemistry, and often also physical chemistry and biochemistry in his premedical collegiate years.

Many undergraduates seeking medical school admission (approximately 11% of total applicants) have concentrated in some area of chemistry. Furthermore, a higher percentage of chemistry majors (43%) are successful in achieving medical school admission than are applicants from the more traditional biology major (34%). In addition, chemistry majors find that they can prepare themselves for careers in biomedical science by pursuing graduate work in biochemistry (approximately 156 schools offer such training in the U.S.A.), in pharmacology (approximately 60 schools), in medicinal chemistry (approximately 33 schools), and in combined Ph.D.-M.D. programs (approximately 40 schools).

Chemistry's contributions to medicine are truly multifaceted. As a discipline, chemistry has provided training to the practitioners of medicine and has supplied them with drugs, clinical laboratory methods, auxiliary polymeric materials, and radioisotopic diagnostic procedures. The interaction of chemistry and medicine has been fruitful in the past; it will no doubt continue to flourish in the future.

ACKNOWLEDGEMENT

The support of the Elsa U. Pardee Foundation is gratefully acknowledged.

SUGGESTED READINGS

BOOKS

Chemistry and the Needs of Society. Special Publication No. 26. London: The Chemical Society, 1974.

Chemistry in Medicine — The Legacy and the Responsibility. Washington, D.C.: American Chemical Society, 1977.

Clarke, F. H. *How Modern Medicines Are Developed*. Mt. Kisco, N.Y.: Futura Publishing, 1973.

Heindel, N. D., H. D. Burns, T. Honda and L. W. Brady. *The Chemistry of Radiopharmaceuticals*. New York: Masson, 1978.

Taylor, J. B., and P. D. Kennewell. *Introductory Medicinal Chemistry*. New York: John Wiley & Sons, 1981.

ARTICLES

Burger, A. "A Quarter Century of Achievement in Medicinal Chemistry." *Chemtech* 5, No. 9(September 1975): 523-528.

Melville, R. S. "Future Trends in Clinical Chemistry." *Analytical Chemistry* 49, No. 7 (June 1977): 594A-597A.

CHEMISTRY AND YOUR FUTURE

Career Opportunities

Carl Pacifico

Carl Pacifico, Senior Partner of Management Supplements, served as President of G & O Mfg. Co. (1971-1972), as Executive Vice President of Alcolac Chemical Co. (1954-1966), and as Director of Development of Wyandotte Chemical Co. (1949-1954). He is the author of Practical Industrial Management, *Wiley, 1981;* Effective Profit Techniques for Managers, *Prentice-Hall, 1967; and* Creative Thinking in Practice, *Noyes, 1966.*

The complexity of the chemical industry results in a wide variety of interesting and professionally satisfying occupations. To provide a fuller understanding of the nature of employment in the chemical industry, its scope, products, contribution to the economy, current problems, and future potential will first be discussed briefly.

Reviewing new chemical developments in technical journals is as thrilling as examining an assortment of sparkling jewels and at least as awesome as contemplating the creative achievements of great artists. Yet chemicals also are practical products to be used, playing a critical role in the international economy; they improve our standard of living as well as enrich our lives in countless ways. The uses of chemicals are so varied that they cannot be contained in the usual definition of a single industry. Moreover, their effect in improving the quality of life extends well beyond the usual numeric measurements of economic value.

There is, of course, a core industry that produces basic chemicals, both organic and inorganic. Many of these products are prepared from natural raw materials which are themselves chemicals though not commonly regarded as such by laymen.

Salt is converted to chlorine and soda ash; fats and oils are split into glycerine and fatty acids; hydrocarbons are separated out of petroleum; a variety of heterocyclic chemicals are distilled from coal; cellulose and related products are obtained from wood.

Another branch of the chemical industry converts basic chemicals into new forms. This may be as simple as neutralizing a fatty acid to make a soap or as complex as multi-step processes to make pharmaceuticals, pesticides, or perfume oils. An especially interesting activity involves linking one or more types of simple units ("monomers") into larger aggregations ("polymers") that are useful as synthetic fibers, plastics, resins, or rubbers.

An important distinction among chemicals, which influences the nature of several employment activities, is that between "composition" and "performance" products. Composition chemicals are those that are used primarily for chemical reactions based on their molecular configurations. Examples are sodium hydroxide, ethylene oxide, and vinyl chloride. Performance chemicals are used exclusively for their physical properties rather than as reactants. Examples are chemicals that are used as lubricants, pigments, solvents, plasticizers (to convert otherwise brittle plastics to useful forms), and catalysts.

Some segments of the chemical industry conduct no chemical reactions but prepare many useful products by simple physical blending of composition and performance chemicals to form paints, inks, laundry detergents, hair shampoos, and similar products that are used for their physical properties. These formulated products will be referred to here as "performance chemical systems."

However, chemicals are indispensible in several activities that are not usually considered to be part of the chemical industry. For example, the textile industry is considered a mechanical industry because it starts with fibers and physically converts them into fabrics. Nevertheless, the modern textile industry could not exist without chemicals used to clean the fibers, lubricate or strengthen them for the weaving operation, clean the fabric after it has been woven, and dye, soften, or waterproof the finished product. Indeed, in some cases the fabrics and rugs are no longer woven but held together by chemical adhesives. A similar situation exists in the paper industry where today the final product may contain a higher percentage of added chemicals than of cellulose itself. Convenience foods could not exist without chemical additives. Many fuels include chemicals to permit more efficient combustion. Asphalt for road construction and roofing is not only a performance chemical itself but is enhanced by solvents and emulsifiers. Concrete is not only formed by a chemical reaction but includes large quantities of chemicals to retard setting time and to entrain air.

Even this extensive list does not cover all the industries where chemistry is sufficiently important so that professional chemists or chemical engineers are required. Several ores are separated and recovered through the use of chemical flotation agents. Chemists are employed to analyze and/or control chemical activities in the manufacture of steel and other metals, in foundries, in municipal water treatment, in the electronics industry, and in the production of business machines. To summarize, there are few aspects of the economy of modern industrial society that do not involve chemicals to some extent.

Common usage limits the term "the chemical process industries" to the following economic activities: industrial inorganic chemicals, industrial organic chemicals, agricultural chemicals, cosmetics and toiletries, drugs and pharmaceuticals, paints and allied products, plastics, soaps and cleaning products, and miscellaneous chemical products. That definition specifically excludes coal products, petroleum, rubber, and plastic processing.

The chemical process industries had sales in 1981 of almost $182 billion, of which individual chemicals represented about $64 billion. During the 1970s, these industries grew at about twice the rate of the average for all U.S. industry, but that is expected to decline to about 1.5 times in the 1980s. About $130 billion, roughly 70%, of those total sales came from the hundred largest chemical producers, including petroleum companies and others which have substantial chemical activities but which are not considered to be chemical companies. There are a few hundred chemical manufacturers with sales in the range of $50-200 million annually. Then there are about 10,000 small chemical companies that comprise the rest of the industry, including individual entrepreneurs making only a few products for local sale.

The chemical process industries in 1981 had about 1.1 million employees, thus representing about 5% of all industrial employment. Of these chemical employees, roughly 60% are factory production workers, compared with 70% for industry in general. The average of sales per employee is now about $110,000. This helps the chemical industry to pay somewhat higher wages to its hourly employees, typically 10-15% more than the average for all industry. As might be expected in a highly technical industry, the chemical process industries employ a high ratio of salaried scientists and engineers. There are about 53,000 technically trained individuals in the chemical process industries, or 4.8% of all employees, compared with 2.3% for all industry.

The chemical process industries make a notable positive contribution to the U.S. trade balance, which has been negative for many years. Chemical exports are typically about twice those of other imports with a resulting net contribution of over $11 billion in 1981.

The more common occupations for chemists and chemical engineers within the industries producing or using chemicals are:

- Exploratory Research — Investigation of the basic mechanism in the performance of chemical systems.
- Synthesis Research — Investigation of specific chemical systems to identify optimum conditions for preparation of a desired product in the laboratory.
- Formulations Research — Investigation of physical combinations of chemicals to develop products with the desired performance properties.
- Applications Research — Identification of the preferred condition of use of the company's products in appropriate industrial applications.
- Product Development — Investigation in the laboratory of modifications by chemical or physical changes that would make a company's products more suitable for their intended industrial uses or for totally new uses.

- Process Development — Translation of laboratory results into commercial equipment, taking into account additional variables such as heat and material transfer, equipment capabilities, and costs.
- Design Engineering — Conversion of the process development data into a design for a workable commercial plant, including selection of appropriate equipment.
- Plant Engineering — Responsibility for the technical aspects of an operating plant, usually including equipment modifications, maintenance, safety, and materials handling and storage.
- Production — Manufacture of product according to product specifications, volume requirements, and shipping schedules at acceptable costs.
- Methods Development — Identification (and occasionally implementation) of methods of analysis to support research activities. Establishment of both procedures and specific tests to identify the quality of plant products before, during, and after manufacture.
- Quality Control — Responsibility for carrying out the test procedures established above on raw materials and finished products.
- Commercial Development — Identification of opportunities for profitable new business activities, including new products, new markets, or acquisition prospects.
- Marketing Research — Responsibility for gathering meaningful commercial information from published sources or by personal field work to assist marketing decisions.
- Market Development — Introduction of new products or established products for new uses or new markets to identify optimum specifications, pricing, conditions of use, and channels of distribution.
- Sales — Representation of a company in inducing potential users to purchase its products.
- Product Management — Responsibility for all aspects of sales support to maximize profit.
- Technical Service — Assistance to customers with technical problems that may arise in storing, handling, or using the company's products.

- Purchasing — Identification of suitable suppliers of the technical raw materials required by the company and negotiation of purchases at acceptable prices.
- Patent Attorney — Securement of patent protection for the company's inventions in appropriate countries. Guidance of activities away from areas patented by others. Negotiation of licenses.
- Safety Engineer — Supervision of storage, handling, and use of all materials to minimize hazards. May also be responsible for legal disposal of all waste products.

Each of these positions involves a combination of technical and interpersonal skills. The duties of a position will vary somewhat from company to company but, in general, the balance of skills required for success in each type of job will be about as shown in the following matrix.

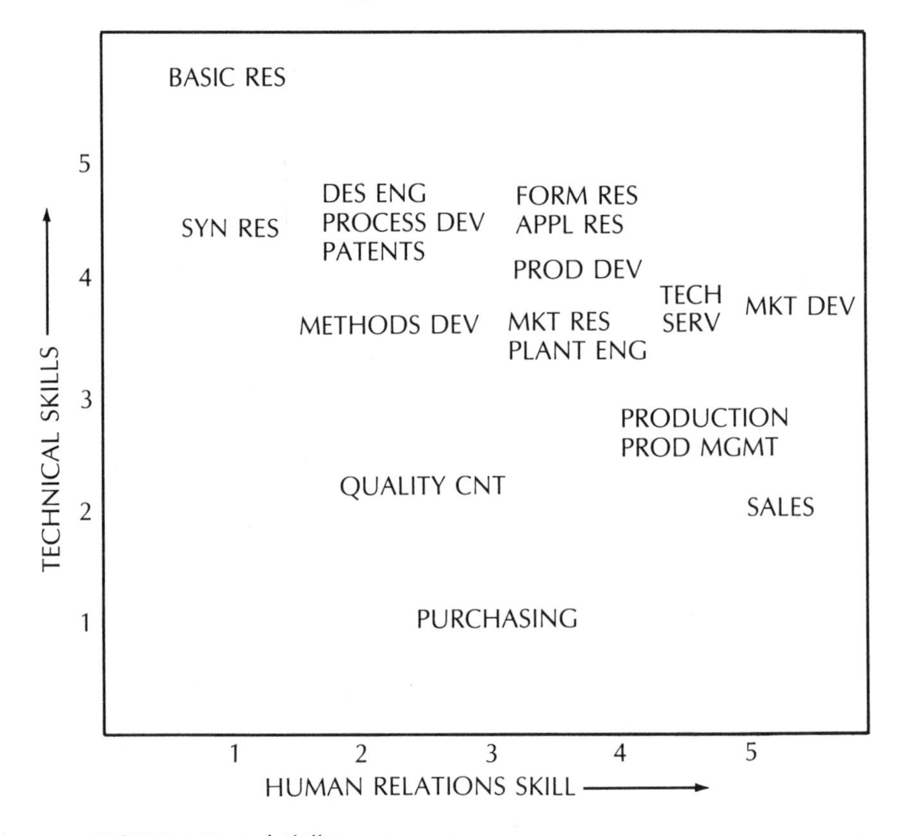

FIGURE 1: Typical Skill Requirements

The personal factors that facilitate success in the chemical industry are about the same as in all other industries. However, educational requirements have become formidable obstacles to progress even for competent individuals who do not have the necessary credentials. Progress beyond the technician level is almost impossible without at least a bachelor's degree. That degree will admit an individual into basic or synthesis research as an assistant, but progress to a supervisory level in those areas is rare without a doctorate (Ph.D.). The Ph.D. is helpful but not universally required for supervisors in applications and formulations research. Advanced degrees are less common in engineering, although a master's degree is helpful in process development and design engineering. In field selling, a graduate degree is not required, but a master of business administration degree is useful for the marketing positions. A patent attorney obviously needs a law degree, but a bachelor's degree in a science is also desirable. A bachelor's degree is adequate for quality control, technical service, and purchasing.

Until rather recently, many competent individuals without degrees, products of an earlier generation when advanced degrees were less common, were able to advance to senior management. However, an increasing number of senior executives of chemical companies now have graduate degrees of Master of Business Administration, doctorate, or both.

An equally serious obstacle to advancement is inadequate or improper preparation for some of the positions described. It is doubtful if there is a formal course offered anywhere in formulating a hair shampoo or an agricultural emulsion. There is little, if any, training in the relationship between the molecular structure of a chemical and its physical properties. No one teaches applications research or technical service techniques. Engineers are taught unit operations and plant design but not practical process development or how to handle union grievances. While a few colleges have recently offered courses in technical sales and related marketing activities, the associations in those fields have been obliged to offer their own training programs. In general, practically nothing is done to prepare students for employment in the performance segment of the chemical and allied industries, which employ at least half of all industrial chemists and engineers.

Despite its great contributions to the economic and personal well-being of the country, the chemical industry has not been free from public criticism. The performance of any industry is influenced significantly by its public image. When an industry is regarded favorably, public policies foster its development, the capital needed for growth is provided by investors at reasonable cost, and competent individuals seek to enter the field and nourish its growth with their talents. A negative public image, on the other hand, has the opposite effect on an industry and retards its progress. Many factors are involved in the public's image of an industry, some of which are valid while others may be misunderstandings or incorrect impressions. In any case, it is important to assess that image realistically in order to make the adjustments needed to improve it.

The public image of the chemical industry has changed considerably during this century. In the first decade, the infant chemical industry was hardly noticed. It made necessary but unexciting products such as caustics, paints, and explosives; almost all the more complex chemicals were imported from Europe. When that source was cut off during the First World War, the chemical industry began production of synthetic organic chemicals, notably dyes. This led to a succession of dazzling achievements of direct value to the public over the next 40 years. The first plastics were introduced, progressing from simple bottle caps to components in a host of household products. Clear films that revolutionized packaging were developed. The wonder drugs, sulfas and penicillins, reduced many infections from life-threatening terrors to minor annoyances. Synthetic soaps minimized the problems of hard water. Then came new paints that could be applied easily and which dried quickly. A series of synthetic fibers followed that wore better than natural materials and were easier to clean. New rubber products gave longer tire life than the natural product, and in the 1950s the foamed products based on polyurethanes were introduced. Throughout this period, the prestige of the chemical industry soared.

Then, for reasons still being debated, the succession of spectacular products stopped. Innovation continued, of course, but now it was directed at product improvements, cost reductions, and industrial uses which were less visible to the public. The period of chemical wonders appeared to be over, and, besides, the public's attention was now attracted to the new wonders of electronics in television, calculators, and computers. For the chemical industry, this proved to be a decade of transition from public approval through indifference to incipient hostility. What the public now saw was industrial fumes, contaminated waters, dangerous fires, and derailed cars that required evacuation of homes. Health and safety became important public issues, heightened by the fear that the industry was harmful to human life. Despite statistics clearly demonstrating the increasing human life span and the reduction of many illnesses, all industry was charged with adversely affecting the quality of life. The chemical industry was further saddled with the hideous accusation that it was the source of many cancers.

Whether or not water and air pollution do represent serious health hazards, they certainly cause the environment to deteriorate and so need to be reduced to acceptable limits. The majority of chemical companies, together with most other industries, responded promptly and responsibly to divert the capital and personnel resources needed to make the required changes in waste disposal as the standards of acceptable performance changed. Unfortunately, a few exceptions among major companies and some really bad examples of illegal disposal by smaller companies have hurt the industry's image by directing attention from its positive accomplishments. In this author's opinion, the chemical industry could have done a much better job, both in policing itself and in its public relations in this area. Nevertheless, substantial progress has been made so that it can be confidently predicted that all targets will be met in the next decade or so. That accomplishment will then permit us to return those resources to the development of new chemicals.

Safety of both employees and the public has always been a concern of the chemical industry. Some chemicals are extremely reactive materials, capable of causing fires and explosions. However, the properties of these chemicals are well known so that equipment and procedures have been developed by the industry to minimize that risk. As a result, the chemical industry has an excellent record, far better than the average of all industry. So, while careful attention to that aspect of safety is continually required, it is not now a serious problem.

In recent years, however, the concept of risks due to chemicals has changed dramatically. The focus now is on the longer-term effect of chemicals on the human body, particularly with regard to cancers. Studies have shown that when mortality data are adjusted for the increasing age of our population, the incidence of cancer (except lung cancer, which is correlated with cigarette smoking) has remained static. Nevertheless, the dread of this disease has resulted in a frantic search for its cause. In some quarters, the chemical industry has been indicted and faced with the difficult problem of establishing its innocence by proving a negative. Nevertheless, reputable analysts have shown that industrial sources cannot be responsible for more than a small fraction of all cancers. Also, most chemical companies have removed all suspected carcinogens from their product lines and are cooperating with new laws requiring tests for carcinogenicity of all new chemicals. As progress is now being made on identifying the true causes of cancer, it can be expected that within the next two decades, the problem will be solved. The chemical industry will then eliminate the few products, if any, that are carcinogenic and so will remove the last obstacle to regaining its favorable public image.

The outlook for the chemical industry over the rest of the century is therefore quite favorable, though it is unlikely to equal the glory period of the mid-century. The industry will continue to grow, though at a lower rate, and make its contribution to the national economy and the welfare of our people, while providing satisfying employment for both scientists and factory workers. Its public image will gradually improve, the resources directed to improving the environment will be returned to growth activities, and our success will again rest on the talents of our executives and the ingenuity of our scientists and engineers.

CHEMISTRY AND CONSUMER GOODS

From Food and Shelter to Sports

Raymond B. Seymour

Raymond B. Seymour, Distinguished Professor of Polymer Science at the University of Southern Mississippi, Hattiesburg, is the author or editor of 25 books and over one thousand research and technical publications. He has been awarded over 45 patents by the U.S. Patent Office.

This prolific author received the BS and MS degrees in chemical engineering and chemistry, respectively, from the University of New Hampshire and the Ph.D. from the University of Iowa. In addition to spending 20 years as a researcher and executive in the American chemical industry, Dr. Seymour has spent 30 years as an educator in American universities. He has also served as a visiting scientist in Australia, Bangladesh, Canada, Czechoslovakia, Taiwan, Trinidad, Yugoslavia, and the USSR. He has served on the editorial boards and as a columnist for Modern Plastics, Plastics, Popular Plastics, Plastics and Rubber, *and the American Chemical Society's* Advances in Chemistry Series.

He is the recipient of the following awards: Western Plastics Award, Excellence in Teaching Award from the University of Houston, American Chemical Society Southeastern Texas Award, Society of the Plastics Industry Plastics Pioneer Award, Outstanding Educator of America, Chemical Manufacturers Association Award for Excellence in College Chemistry Teaching, American Institute of Chemists Honor Scroll, Western Plastics Hall of Fame, Society of Plastics Engineers International Education Award, and Southern Chemist Award.

Unlike primitive man, who was able to supply his basic needs without benefit of science, modern man must rely, to a large extent, on science and technology for his food, clothing, shelter, exploration, recreation, ventilation, communication, and education.

AGRICULTURE

Early man obtained food by hunting, fishing, and gathering nuts, berries, and fruit. However, as he became more civilized, he domesticated wild animals and cultivated wild crops such as millet, rye, barley, rice, and wheat. As a result of the invention of simple tools, such as the hoe, scythe, and plow, man was able to develop the crude science of agriculture and to increase the yield and quality of his crops.

Jeffrow Wood's invention of the iron plow with standard interchangeable parts in 1819 catalyzed a new era in agriculture. The subsequent development of machines such as the reaper, mechanical harvester, combines, and tractors made possible modern agronomy, i.e., the large scale production of staple crops.

In an essay on the "principles of population" published in 1798, Thomas Malthus predicted that since populations increase much faster than the development of food supplies, worldwide starvation would occur in the near future. Fortunately, advances in agricultural chemistry and contributions by agronomists such as Nobel Laureate N. E. Borlaug have extended the time before catastrophic starvation by the development of superior grains and the use of irrigation, fertilizers, and pesticides.

European farmers used animal manure and the American Indians used dead fish as fertilizers to supply nitrogen to the soil. Later, guano deposits and the Chilean sodium nitrate deposits were used on a large scale. Some nitrogen is also supplied to the soil by legumes which convert nitrogen from the air to nitrates in a process called fixation. Unfortunately, all of these sources of nitrogen are insufficient for today's agricultural demands.

Small amounts of ammonia, called "spirits of hartshorn," were obtained by the alchemists through the destructive distillation of deer horns (hartshorn). However, in 1912, Haber discovered the fixation of nitrogen process which is widely used today for the production of ammonia. In this process, nitrogen (N_2)

from the air is combined with hydrogen (H_2) under pressure and in the presence of metallic oxide catalysts to yield gaseous ammonia (NH_3).

The ammonia gas is dissolved in water and used directly as fertilizer; it is also catalytically oxidized to nitric acid (HNO_3). Thus a solid fertilizer, ammonium nitrate (NH_4NO_3), can be made from ammonia and nitric acid, and many other solid ammonium and nitrate salts can be produced from the reaction of these and other reagents.

Potassium (K) salts (potash) are mined at Trona, California, and Saskatchewan, Canada. Insoluble calcium phosphate ($Ca_3(PO_4)_2$) is mined in Florida and Tennessee and converted to a soluble product (superphosphate) by treating it with sulfuric acid (H_2SO_4). The monobasic calcium phosphate ($Ca(H_2PO_4)_2 \cdot H_2O$) is soluble in water.

The three most common components of commercial fertilizers, viz., nitrogen (N), phosphorus pentoxide (P_2O_5), and potassium oxide (potash, K_2O), are listed in that order on each bag of fertilizer. Thus, a label of 10-10-5 indicates a nitrogen, phosphorus, and potassium content of approximately 10, 10, and 5 percent, respectively. One may also use the following mnemonic expression to remember the essential nutrients: "C. F. Hopkins' CaFe where there is NaCl on the table and the food, which is cooked by Maney (Mn) Molly (Mo) whose cuisine (Cu) is mighty good (Mg) and there are no dirty dishes in the sink (Zn)."

In addition to fertilizer, the chemist must also supply pesticides which allow the crops to be harvested before they are destroyed by insects. While parasites and predators are biological pesticides, most agricultural pesticides are contact, stomach, or fumigant poisons which are applied as sprays, gases, or seed coatings.

In 1948 the Swiss chemist Paul H. Müller received the Nobel Prize in Physiology or Medicine for his discovery that dichlorodiphenyltrichloroethane (DDT) is an effective contact insecticide. Millions of lives were saved by the application of this insecticide during World War II. The entire population of *Anopheles* mosquitoes, which transmit malaria, was almost annihilated before the use of DDT was abolished.

DDT and many other chlorinated insecticides are not specific for any one insect. They kill both good insects and pests, are soluble in the fatty tissue of birds and animals and are not readily decomposed. Hence, the use of these so-called "hard insecticides" is now severely restricted. Because of previous widespread and sometimes careless use of DDT, many individuals and large birds have a residue of as much as 20 parts per million (ppm) of this insecticide in their body fat.

The most widely used insecticides today are the organophosphates, such as methyl parathion and malathion. These organophosphate pesticides are relatively complex compounds containing both phosphorus (P) and sulfur (S) and are derivatives of phosphoric acid (H_3PO_4). These insecticides are more toxic than the chlorinated pesticides, like DDT, but remain active for a shorter period of time and thus are not accumulated in the food chain.

Carbamates, such as carbaryl, are less toxic than the organophosphates and also have a short residual activity. These aromatic compounds, related to urea (H_2NCONH_2), contain benzene (C_6H_6), which is a cyclic hydrocarbon, i.e., the carbon atoms are joined together to form a hexagonal ring structure

```
          H
        / C \\
     HC       CH
     ||       |
     HC       CH
        \ C //
          H
```

Paradichlorobenzene (ClC_6H_4Cl) is a good example of a fumigant-type insecticide. This solid compound vaporizes in the atmosphere and is widely used as a mothicide. Soil fumigants, such as ethylene dibromide ($C_2H_4Br_2$), are particularly effective for the control of nematodes (roundworms).

Inorganic compounds such as sulfur (S) and compounds of copper (Cu) which have been used as fungicides are being replaced by chlorinated aromatic compounds. Carbon-containing compounds are called organic compounds, while those containing the other elements are called inorganic compounds.

As a result of the use of pesticides, a farm worker is able to supply twice as much food today as he could 10 years ago.

Other applications of science to agriculture have increased the farm worker's efficiency 1000 percent during a period of 50 years.

While over four million tons of pesticides are used annually, worldwide, there is a good possibility that their use will decrease because of other pest control techniques such as the use of pheromones or sex attractants. Female insects emit as little as 10^{-15} grams (0.000000000000001 g.) of pheromones. The latter are usually specific unsaturated hydrocarbons which can be snythesized commercially. The specific pheromones can be used to attract and trap undesirable male insects or to attract other insects which will destroy the pests.

To illustrate, the pheromone in the female cotton leaf worm has the chemical structure

$$CH_3\text{-}CH_2\text{-}CH{=}CH\text{-}CH{=}CH(CH_2)_8\text{-}O\text{-}\underset{\underset{\displaystyle O}{\|}}{C}\text{-}CH_3$$

CLOTHING

Animal skins served as ancient man and woman's clothing, but there would not be enough animals available today to supply such essential needs. When man learned the value of agricultural crops and how to keep domestic animals, he was able to make crude garments from flax, wool, silk, and cotton, and by trial and error, he discovered how to clean, scour, and dye these articles of clothing.

His first cleansers were simply the extracts of wood ashes (potash), but later he heated this alkaline material with animal fat and obtained soap. The original dyes, which were vegetable colors, were replaced by synthetic dyes after William Perkin discovered in 1856 how to make a purple dye, called mauve, from aniline. The modern chemist not only supplies thousands of synthetic dyes with many colors but also provides synthetic detergents which do not leave a ring around the bathtub or collar. All detergents are able to attract water and grease and to disperse the latter in the former because of the presence of water-loving (hydrophilic) and grease-loving (lyophilic) groups in the same molecule.

Flax or *Linum usitatissimum* is a cellulosic fiber which was the major source of cloth before the nineteenth century. This plant has been cultivated for its fiber and seed since prehistoric times. The seeds are crushed to make linseed oil, an unsaturated oil, which is used as a drying oil in oleoresinous paints.

The fiber is obtained by allowing bacteria to break down wetted woody tissues in a process called retting. The fibers are then separated from the pith by beating (scratching) and combing (hackling). The linen yarn obtained from the flax fiber is woven to produce linen cloth. The linen textile industry was established in the seventeenth century near Belfast, Ireland. Ireland continues to be the world's leading producer of linen.

Wool is a proteinaceous fiber obtained since prehistoric times by shearing domesticated sheep. George Washington imported sheep, spinners, and weavers from England, which was the major wool-producing country in the eighteenth century. The first American woolen mill was built at Hartford, Connecticut in 1788, and New England has continued to be the leading area for the production of woolen textiles.

The fiber-making process consists of carding (combing), drawing the fiber into roving or unspun yarn, and then spinning the scoured and dyed wool fiber. About 2.7 million tons of wool are produced annually, with most of this production in Australia and New Zealand.

Silk is also a proteinaceous fiber which is spun by the larva of the *Bombyx mori* or mulberry silkworm. The culture of the silkworm (sericulture), the reeling of the continuous filament from the cocoon, and the weaving of silk has been practiced in China since about 2640 BC.

Silkworm eggs and mulberry tree seeds were smuggled from China to Turkey by missionaries about A.D. 550. Silk weaving later was gradually moved to western Europe. Ben Franklin fostered a silk industry in Pennsylvania, using silk from mulberry trees in Georgia. However, this labor-intensive industry did not thrive in the U.S., and Japan and China continue to be the principal suppliers of silk. This expensive natural fiber has been displaced, to a large extent, by less expensive man-made nylon.

Cotton, which belongs to the genus *Gossypium* and to the families *mallvacea, hersutum*, and *herbacem*, is the most important natural fiber. This cellulosic fiber has been spun, woven, and dyed in Egypt, India, and China since prehistoric times. Cotton cloth was also produced in Peru at least 1000 years ago.

Labor-intensive cultivation of cotton began in Jamestown, Virginia in 1607. The tedious process of seed separation (ginning) was mechanized by Eli Whitney's invention of the cotton gin in 1793. The southern states in the USA soon became the world's leading producers of cotton, and the mills in New England produced a large share of woven cotton textiles. Many of these production facilities were moved to the southeastern United States in the twentieth century.

Cotton seeds, which contain about 20% oil, are crushed for their oil. Much of this unsaturated oil is converted to a solid by the catalytic addition of hydrogen (H_2). The unsaturated oil may be used in paint manufacture and as a cooking oil. The hydrogenated oil is used for margarine and shortening (Crisco, Spry, etc.).

Like other natural fibers, cotton fibers are carded, combed, and spun to produce cloth. Because of the large-scale production of man-made fibers, cotton is no longer king. However, more than 14 million tons of cotton are produced annually worldwide.

With the exception of silk, which is spun by moths as a continuous filament, all other natural fibers are short or staple fibers. In 1655 Robert Hooke suggested that silk could be made "artificially," but the first man-made filament was not produced until more than two centuries later when, in 1884, Count Hilaire de Chardonnet forced a collodion solution through small holes (spinnerets) and produced a continuous filament by evaporation of the solvent. Collodion is produced by dissolving cellulose nitrate in a mixture of ethyl alcohol (ethanol) and ethyl ether.

Chardonnet's "artificial"silk was displaced by the less flammable rayon made by the viscose process in 1892. Viscose or the sodium salt of cellulose xanthate is produced by the reaction of carbon disulfide (CS_2) with cellulose in the presence of sodium hydroxide (NaOH). The xanthate filaments are converted to cellulose when they are passed through an acidic solution. Cellophane is similar to rayon, but in this case the cellulose xanthate is forced through slit dies before being precipitated in an acid bath. More than three and a half million tons of rayon are produced annually worldwide, but this volume is decreasing because of competition from synthetic fibers. The production of cellophane is also decreasing for the same reason.

Acetate rayon is produced by passing an acetone (CH_3COCH_3) solution of cellulose diacetate through spinnerets and evaporating the solvent. The cellulose triacetate is produced by reacting cellulose with acetic anhydride $(CH_3CO)_2O$ in the presence of sulfuric acid. The triacetate is then heated with a solution of sodium hydroxide in order to remove about one third of the acetyl groups (CH_3CO-) and produce an acetone-soluble cellulose diacetate. Acetate rayon fibers have poor strength but have a desirable feel or "hand."

The first truly synthetic fiber was nylon, which was synthesized by W. H. Carothers of E. I. duPont de Nemours in the 1930s. There are several commercial nylons, but the original synthetic fiber, nylon-66, which is produced at an annual rate of about one-and-a-quarter million tons in the US, is the major nylon fiber.

Nylon-66 is produced by heating a salt formed by the reaction of a difunctional acid (adipic acid, $HOOC(CH_2)_4COOH$) and a difunctional amine (hexamethylenediamine, $H_2N(CH_2)_6NH_2$). The polyamide contains the repeating unit of

$$(-NH(CH_2)_6NHCO(CH_2)_4CO-)$$

This polyamide is heated and the molten product is forced through spinnerets in a process called melt spinning. Nylon-66 will consist of at least 100 repeating units in the polymer molecule.

The strength or tenacity of nylon-66 and of other extruded filaments is increased by stretching or drawing. The drawing process allows the long polymer chains to slip by each other until they find an attractive unit in an adjacent chain. Much of the strength in nylon and silk is the result of hydrogen bonding, i.e., an attraction between the carbonyl (C=O) groups in one polymer chain and the hydrogen atoms of the amide groups in an adjacent chain:

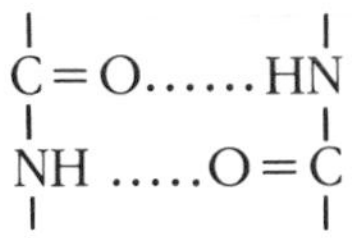

```
 |          |
 C=O......HN
 |          |
 NH .....O=C
 |          |
```

The properties of the various nylons are dependent on the number of methylene (CH_2) groups in the repeating unit. Qiana, a nylon produced by the condensation of dodecandioic acid (a 12-carbon acid) and di-(4-aminocyclohexyl)methane (a cyclic, ring-structured diamine), has a feel (hand) and draping quality like silk.

The most widely used synthetic fiber is a polyester (polyethylene terephthalate) produced by the reaction of ethylene glycol, $HO(CH_2)_2OH$, and terephthalic acid, $HO_2C\text{-}C_6H_4\text{-}CO_2H$, or its derivatives. The polyester filaments are obtained by melt spinning of the molten polymer. Polyester fibers are used for the production of double-knit fabric and are also blended with cotton to produce wash-and-wear wrinkle-resistant garments. Garments made from polyester fibers which have low moisture absorption have a clammy feeling, but the cotton in the blend absorbs moisture and overcomes this undesirable effect. Over three million tons of polyesters are produced annually in the US. This polyester is also formed into pipe-like sections that are air-blown to produce bottles or films.

Acrylic fibers, which are the third most important man-made fibers, are produced at an annual rate of 350 thousand tons in the US. These fibers are produced by the polymerization of acrylonitrile ($CH_2=CHCN$) and are obtained by passing a solution of the polymer in dimethylacetamide through spinnerets and evaporating the solvent.

Polypropylene, with the repeating unit of ($\text{-}CH_2\text{-}CHCH_3\text{-}$), is melt-extruded through a slit die to produce a film which is fibrillated to form fibers. In the fibrillation process, warmed twisted film strips or tape are stretched and passed over closely spaced pins. The polypropylene fibers produced in this way are used for the manufacture of indoor-outdoor carpets and rope.

A specialty elastic fiber, called Spandex, is based on a copolymer consisting of multiple urethane repeating units, ($\text{-}(CH_2)_3NHCOO\text{-}$) and multiple flexible polyester units,

$$(\text{-O}(\text{CH}_2)_2\text{-O-}\overset{\overset{\displaystyle O}{\|}}{\text{C}}(\text{CH}_2)_4\overset{\overset{\displaystyle O}{\|}}{\text{C}}\text{-})$$

The flexible polyester units provide elasticity, which is controlled by the stiffer polyurethane units. These fibers are used in swimwear and foundation garments.

Glass fibers are produced by passing molten glass through spinnerets and are used as reinforcements for plastic composites. Fibers of Kevlar, which is a nylon-like aromatic polyamide or aramid, and graphite (carbon), are also used for high-performance composites. Kevlar is also used in automobile tires and bullet-proof clothing.

SHELTER

While ancient man was a cave dweller, modern man prefers to live in structures made from wood, brick, Portland cement, or steel. The building of a crude log cabin does not require the services of a chemist, but high-quality bricks, cement, and steel are products of modern chemistry. Likewise, the paint used to beautify and protect the exterior and interior of wooden structures, as well as plastic pipes, siding, counter tops, and panels are all products of modern chemistry.

Prior to the twentieth century, paints were made from natural pigments and vegetable oils like linseed oil, but many modern paints are based on synthetic resins, such as polyvinyl acetate and polymethyl methacrylate. The trend is toward the use of high solids and water-dispersed coatings in order to minimize contamination of the atmosphere by organic solvents.

Polyvinyl acetate is produced by the polymerization of vinyl acetate ($H_2C = CHOOCCH_3$). The polymer is obtained as a milk-like latex with soap used as a dispersant or emulsifier during the polymerization. Polymethyl methacrylate latex is obtained in a similar process by the emulsion polymerization of methyl methacrylate ($H_2C{=}C(CH_3)COOCH_3$). The water-borne paint is obtained by the addition of dispersions of pigments to these emulsions.

Clear sheets of polymethyl methacrylate (Lucite® or Plexiglas®) may be produced by forcing the dry polymer through a slit die in an extrusion process which is similar to that used to make spaghetti. The clear plastic sheets are used in place of glass when resistance to breakage is required.

Polyvinyl chloride (PVC), which is produced by the polymerization of vinyl chloride ($CH_2 = CHCl$), is extruded to produce pipe, siding, floor tile and window frames. Counter tops and panels may be produced by building up layers of paper impregnated with phenol-formaldehyde resin and laminating this system by hot pressing. Since the phenolic resin is dark in color, it is customary to use paper impregnated with melamine-formaldehyde resin for the top layer. Cellulose-filled melamine resins are also molded to obtain dishware.

Other polymers, such as polystyrene and polyethylene, are used for disposable dishware and packaging, respectively. Foamed polystyrene is also used as an insulating material, and extruded polyethylene is used as a flexible tubing. These plastics are produced by the polymerization of styrene, $CH_2{=}CH(C_6H_5)$, and ethylene, $CH_2{=}CH_2$, respectively.

EXPLORATION

Ancient man increased his exploratory powers by domesticating animals, such as horses, donkeys, and camels, and by building rafts and boats. This activity was improved dramatically by the use of power derived from the combustion of coal or gasoline. Air travel was made possible by use of petroleum hydrocarbon fuels and jet engines. The high thrust from burning fuels in rockets has permitted the astronauts to visit the moon and space ships to land. While little synthetic chemistry is involved in horseback riding or sailing, considerable chemistry is involved in the production and combustion of modern fuels used in automobiles, airplanes, rockets and the components of life-support systems.

For example, rubber used for tires and flexible parts of automobiles demonstrates the role of chemistry. Much of the rubber in automobiles and truck tires continues to be natural rubber (*Hevea brasiliensis*), but this traditional product is being displaced to some extent by synthetic elastomers.

Natural rubber was of little use until Charles Goodyear cured or vulcanized it by heating it with small amounts of sulfur in

1838. Rubber in the modern tire is reinforced by carbon black (C) and is stabilized by antioxidants which prevent deterioration during outdoor exposure.

A synthetic elastomer, called *cis*-polyisoprene, contains the same isoprene ($CH_2=C(CH_3)CH=CH_2$) building block found in natural rubber. The building blocks in oil-resistant neoprene rubber is chloroprene, $CH_2=CClCH=CH_2$.

The most widely used synthetic rubber is a copolymer of butadiene, $CH_2=CHCH=CH_2$, and styrene, $CH_2=CH(C_6H_5)$. This SBR rubber contains both butadiene and styrene building units in its polymer chain. Other specialty elastomers are copolymers of ethylene and propylene and copolymers of isobutylene, $CH_2=C(CH_3)_2$, and isoprene (butyl rubber). It is customary to cure or cross link these polymers by forming some chemical bonds between adjacent polymer chains so that they will not flow and form flat spots on the bottom side of the tires.

VENTILATION, HEATING, AND COOLING

While the caveman enjoyed a relatively uniform temperature in his cave, this is not the case with stone, wood or brick houses. Chimneys were constructed to vent the smoke from the wood or coal fires and considerable ingenuity was displayed in attempts to circulate the heat through early buildings. But the cooling required by homes in the sunbelt is made possible by modern chemistry.

One of the more advanced devices is a heat exchanger called a heat pump, which uses a volatile liquid such as Freon® (Cl_2CF_2) to transfer heat from either the interior or exterior, depending on the outdoor temperature. Volatile liquids like Freon® are also used as refrigerants for the preservation of foods.

RECREATION (SPORTS)

One of the advantages associated with advances in civilization is that less time is required to secure the basic needs of food, shelter and clothing, hence more time is available for other activities such as recreation and sports. Primitive games such as lacrosse and even the original golf game required only a crude elastic ball and net or stick. However, the skill and enjoyment of these and other sports have been enhanced as a result of chemistry.

Games like lacrosse and tennis are no longer played with wood racquets equipped with gut strings but rather with racquets manufactured from graphite-reinforced epoxy resins and aramid strings.

The original hand-molded sand golf tees have been replaced by tees that are injection molded from a plastic such as polycarbonate, and golf balls are now produced by molding a synthetic rubber called *cis*-poly-1,4-butadiene, which is superior to natural rubber. The modern golf club shafts, like tennis racquets and fishing poles, are made from graphite-reinforced epoxy resins by means of a "pultrusion process," in which a bundle of filaments impregnated with uncured resin are drawn through a die and cured by heating.

Other contributions of the chemist to sport and recreation are football helmets molded from ABS (acrylonitrile-butadiene-styrene) copolymer, and racing cars and boats of fiberglass-reinforced polyester (GRP) plastics. Rigid unicellular polyurethane and polystyrene foams improve the buoyancy of plastic boats.

COMMUNICATION

The art of printing was essential for the development of modern civilization. It utilizes paper made from sheets of cellulosic fibers and ink which is a dispersion of carbon black and synthetic resins in a volatile solvent. The printed word, as a means of communication, has been reinforced by the use of the telephone, radio, television cameras and all sorts of copying and recording machines. Each of these modern instruments of communication is dependent on chemistry, both for signal transmission and for the hardware used in the construction of the instruments.

For example, modern telephones, radios and televisions would not be possible without the use of plastics as insulators and solid state components. Computers are also products of modern chemistry. Radar was made possible by the use of the coaxial cable based on polyethylene. The latter, unknown in the early 1930s, was synthesized in the laboratory and produced commercially so that radar became a reality in World War II.

DECORATION

The caveman decorated the walls of his cave by suspensions of pigments in oils. The quality and versatility of these coatings was improved as man discovered a larger variety of naturally occurring pigments and the so-called drying oils. The latter solidify in the presence of oxygen from the air by a chemical process called polymerization or cross-linking. Modern paints that "set" or polymerize in a few seconds in the presence of ultraviolet light are now available.

While Henry Ford offered his Model T car in any color as long as it was black, the consumer sought other colors. These were readily provided by the chemist. The original Duco® lacquer was a dispersion of carbon black in a solution of cellulose nitrate. This flammable plastic, which had been used for photographic film and many decorative articles, has been replaced by other polymers such as the alkyds and polymethyl methacrylate. While some natural pigments are still in use, the chemist has increased the decorative qualities of paints by synthesizing many new pigments.

EDUCATION

Slate boards and chalk used in the 19th century have been replaced, to some extent, by paper and plastic ballpoint pens or pultruded plastic pencils. Modern audiovisual aids include cellulose acetate film in projectors and polyester tapes in recorders and duplicators.

While many of these educational aids based on chemistry are important, they are less important perhaps than new curricula in which chemistry is one of the required courses. A scholar of the nineteenth century learned Latin and Greek so that he could understand the language of ancient philosophers who had influenced classical, humanistic education.

However, in this age of science and technology, one cannot be considered properly educated unless one also knows the elements of chemistry. Since most consumer goods are the products of chemistry, a knowledge of chemistry is essential not only for obtaining food, shelter and clothing, but also for an understanding of the world around us.

As consumers, we are dependent on chemistry for the nourishing, unspoiled and tasty food we eat, the safe water we drink, the books we read, the drugs that cure our sickness, the homes that keep us comfortable and even for the air we breathe. Unfortunately, waste products from some of the processes essential for the production of consumer goods have contaminated some of our sources of water and air. The chemist, when called upon, can reduce this pollution and assure a longer and happier life for those who live in this modern chemical age.

SUGGESTED READINGS

Carson, R. *Silent Spring*. Boston: Houghton-Mifflin, 1962.

Chemistry and Life. Edited by W. E. Maxwell. Belmont, Calif.: Dickenson Publishing Co., 1970.

Fisher, C. H. *History of Natural Fibers*. Edited by R. B. Seymour in *History of Polymer Science and Technology*. New York: Marcel Dekker, 1982, pp. 281-311.

Moncrieff, R. W. *Man Made Fibers*. New York: Wiley, 1975.

Morgan, P. W. *Brief History of Fibers from Synthetic Polymers*. Edited by R. B. Seymour in *History of Polymer Science and Technology*. New York: Marcel Dekker, 1982. pp. 49-68.

Seymour, R. B. *Modern Plastics Technology*. Reston, Va.: Reston Publishing Co., 1975.

Stine, W. R. *Chemistry for the Consumer*. Boston: Allyn and Bacon, 1978.

Ucko, D. A. *Living Chemistry*. New York: Academic Press, 1977.

Venugopolan, M. *Chemistry and Our World*. New York: Harper and Row, 1975.

Wittcoff, H. A., and B. C. Reuben. *Industrial Organic Chemicals in Perspective, Part 2, Technology, Formulation and Use*. New York: Wiley-Interscience, 1980.

CHEMISTRY AND FORENSIC SCIENCE

Aiding the Criminal Justice System

James J. Hazdra

James J. Hazdra received his B.S. degree in Chemistry from Illinois Benedictine College in 1955 and his Ph.D. in Chemistry from Purdue University in 1959. He has worked as a research chemist and Director of Product Research in industry. In 1963 he became a Professor of Chemistry at Illinois Benedictine College and served as Chairman of the Department of Chemistry and Biochemistry from 1965-77. In 1977 he became Director of Health Care Education for Illinois Benedictine College. From 1963 to the present time he has been an active consultant in the field of forensic chemistry.

Chemistry is constantly becoming more and more important to the justice system. The total crime rate in our society is increasing, and rapid crime detection can only be accomplished by greater application of scientific technology. Just as the ways and means of criminals are becoming more sophisticated, so the tools now available to the criminalist are also forever expanding, ranging from a single magnifying glass to a scanning electron microscope.

Forensic science can be generally defined as the application of scientific principles to the administration of justice. Forensic chemistry is a branch of forensic science which involves the use of chemical principles and techniques in the examination of all types of materials that are items of evidence.

One of the basics in forensic science is Locard's exchange principle which states that any contact between two objects such as the criminal and something at the scene of a crime cannot occur without a transfer of material between them. Any person entering and exiting an area leaves something behind or takes something with him. It is the task of the criminalist to discover these traces and to assess their significance. However, the practical realization of that principle is often very elusive.

Samples collected in criminal investigations and used as evidence may be as varied as fabrics, glass, metals, drugs, blood, building materials, firearms, discharge residues, cosmetics, documents, explosive residues, paints, rubbers, fibers, hair, flammable liquids and residues, finger and palm prints, inks, intoxicating compounds, saliva, seminal fluid, soils, and urine. All these must be examined by the criminalist using instrumentation ranging from the simple to the highly sophisticated. The forensic chemist must always keep in mind that the results of his examination may assist the criminal investigation and will be used in a court of law to prove the innocence or guilt of individuals.

HISTORICAL DEVELOPMENT OF FORENSIC LABORATORIES

On the morning of February 14, 1929, in Chicago's near north side the infamous St. Valentine's Day Massacre took place. Five of the murdered men were found to be members of the "Bugs" Moran gang. A problem arose for the law enforcement official because it appeared as if the Chicago police had been involved in the massacre. The coroner, Dr. Herman Bundesen, ordered the collection of nearly 70 empty shells, bullets, and bullet fragments. Because of the unusual complexity of this case, a special jury of six men, all prominent in the Chicago community, was formed to study the case. Bret A. Massee, vice-president of Colgate-Palmolive Peet Co., was chairman of this jury. From New York City the jury brought in Calvin Goddard who had gained a national reputation for his firearm work. Goddard quickly concluded that all of the shells and bullets were fired from two .45 caliber Thompson submachine guns. The markings of the collected bullets did not match any of the Thompson machine guns found in the Chicago police department. Almost 11 months later machine guns producing identical markings as those found in the bullets of the Chicago massacre were confiscated from killers in St. Joseph, Missouri.

This set the stage in Chicago for the establishment of the first crime laboratory in the United States. Massee continued his financial support and suggested that Chicago should have a crime laboratory not only for the purpose of identifying firearms but for the application of scientific methods to other brands of criminal investigation. A laboratory corporation was formed and affiliated with Northwestern University in June, 1929. Goddard became the director of the Scientific Crime Detection Laboratory, the first of its kind in the nation. Northwestern's Chicago Laboratory, through its sense of an educational mission, conducted information and training seminars. Attending these classes in 1931 were agents from the now well-known Federal Bureau of Investigation. This resulted in the establishment of an official Criminological Laboratory for the FBI in Washington, D.C. on November 24, 1932. In 1943 the name was changed to the FBI Laboratory, which today is one of the largest and best equipped laboratories of this kind in the United States.

There exist now a little more than 250 crime laboratories in the United States. Each state and almost every major city has one. On the federal level, in addition to the FBI, the Bureau of Alcohol, the Tobacco and Firearms Administration, and the Drug Enforcement Administration also have crime laboratories.

However, the training of manpower has been a problem, especially for the newly organized laboratories. In many cases, several newly developed instrumental techniques have outrun the forensic scientist's ability to utilize them as important tools for crime detection. Much more developmental research is needed in applying all available analytical tools to crime detection.

TOOLS OF THE CRIMINALIST

Crime laboratories have come a long way since the original laboratory in Chicago, growing beyond firearm and fingerprint identification. Summaries of a few major chemical instrumentation methods used in crime laboratories follow. The classical wet chemical analyses (such as spot tests) described by Feigl for identification of both inorganic and organic substances will not be discussed here.

Atomic Absorption Spectrophotometry (AAS) This technique involves the quantized absorption of ultraviolet and visible radiation leading to atomic excitation. It is becoming extremely useful in the analysis of soils, building materials, and alloy metals since it can determine how much of each trace metal is present. The purpose of this analysis is to determine where an evidence sample may have come from or to compare samples.

Recently, the advent of flameless graphite furnace techniques has increased the sensitivity of detection by AAS and has virtually replaced the very expensive and time-consuming method of neutron activation analysis for traces of lead, barium, and antimony from firearms discharge residues. The relatively new AAS techniques have made it possible to reach a concentration range of parts per billion, which may even reach beyond the limits of detection once afforded by neutron activation analysis.

Chromatography This is a process long used by forensic chemists which allows the resolution of mixtures by effecting the separation of some or all of their components into concentration zones or phases different from those in which they are originally present, irrespective of the nature of the force or forces causing the substance to move from one phase or another. Paper and thin-layer chromatography in conjunction with infrared spectrophotometry have been used for drug and metabolite separation and identification. Approximately 30% to 40% of the work load of the forensic chemist involves drug identification. In order to make this job easier gas chromatography-mass spectroscopy (GCMS) has been employed for efficient separation and identification of drugs so as to detect controlled substances. Gas chromatography is a process which separates substances using the gas-liquid phase. Mass spectroscopy, which measures the mass to charge ratio of gaseous ions, can identify various gaseous substances coming off the gas chromatography column.

Since most drugs and cosmetics cannot be volatilized, high performance liquid chromatography with ultraviolet or infrared spectrophotometers as detectors are beginning to be utilized for many fast separations and identifications. In addition to drugs, both quantitative and qualitative analysis of petroleum products, inks, paints, explosive residues, and plastics by use of pyrolytic techniques in GCMS have provided useful information in crime detection. Ultraviolet spectroscopy involves the quantized absorption of ultraviolet electromagnetic radiation leading to specific electronic excitation of molecules. Fluorescence spectroscopy involves the quantized emission of ultraviolet visible electromagnetic radiation resulting from the electronic deexcitation of molecules. Both of these techniques can be used for compound identification as well as determination of the quantity of the compound present. They have been extremely useful in the detection of inks and invisible laundry marks. They routinely are used in preliminary examinations of seminal stains as well as altered or erased documents.

Infrared Spectroscopy This technique involves a quantized absorption of infrared light leading to vibrational excitation of molecules which can be used for compound identification. This method of analysis has been useful in distinguishing pigments and dyes that appear identical to the naked eye as well as in identifying many rubbers, plastics, and other organic compounds.

In the area of building materials, paints, drugs, cosmetics, petroleum products, fibers, etc., infrared and ultraviolet spectrophotometry have been useful in comparing and identifying these substances.

Energy Dispersive X-Ray Fluorescence (EDXRF) This technique is being used by an increasing number of forensic chemists for elementary pattern identification and detection of key indicating elements in order to compare evidence samples. It uses a known X-ray wavelength to interact with a specimen. These X-rays are absorbed by the specimen resulting in the release by the specimen of secondary X-radiation or "flourescence X-rays." The pattern of wavelengths of this fluorescent radiation is identical with that of the X-ray emission spectra of the various elements present. The method is a rapid nondestructive analytical technique, requiring only several minutes, capable of simultaneous multielement analysis of both trace elements in bulk and submicrogram samples (less than one-millionth of a gram) at the parts per million level (ppm). It is easily interfaced with a minicomputer for automated data accumulation, data processing, correlation studies, and comparisons. These unique features of EDXRF have been applied to a wide range of samples from bullet holes to hair and including glass fragments, clothing, drugs, building materials, paints, inks, firearms, and residues. As an example, reexamination of a bullet hole profile by monitoring the intensity of lead, tin, antimony, barium, and copper can produce an X-ray spectrum. Thus it is now possible to establish the distance between the gun and the victim, the angle of the shot, etc.

EDXRF has also been extremely useful in paper, document, and currency examination and analysis. For instance, 18 elements have been detected in a $20 bill, which can be matched qualitatively and quantitatively to identify counterfeit money with paper and ink found on a suspect's premises. New techniques are constantly improving the use of EDXRF, and its use in the forensic analytical laboratory will increase in the future.

X-Ray Diffraction This technique identifies solid compounds by measuring the interference between scattered X-rays caused by the crystal structure of the solid. X-ray diffraction powder pattern techniques have been used for 20 years or more to identify and compare unknown crystalline substances in the forensic laboratory. The X-ray diffraction spectrometer records the manner in which X-rays are bent to identify mixtures of crystals found in building materials, explosive residues, different alloys, soils, etc.

Luminescence Spectroscopy Recently it has been discovered that through the use of luminescence, traces of substances left by contact can be detected by causing these traces to fluoresce or phosphoresce. Luminescence phenomena occur at characteristic wavelengths for different substances. For example, tire prints can be obtained from this fluorescent pattern because of the transfer of tire rubber extender oils on the surface. The intensity of this fluorescence can be used to determine the age of the tire surfaces.

A number of other applications for this technique are obvious, e.g., shoe scuff marks will contain substantially different types of material varying from shoe to shoe, rubber eraser marks from pencils, identification of tire skid marks in a hit-and-run case, etc.

At present, chemical and spectroscopic studies probably show less than 50% of the luminescence energy available from a print that could be used to detect it. The development of image conversion and identification techniques in conjunction with new sources of excitation such as lasers would considerably increase the efficiency of detection.

Analyses of blood and other body fluids are becoming more chemical than biochemical problems. Evidence is collected mainly in connection with crimes of murder, rape, robbery, assault and battery, hit-and-run, or game violations. It is relatively simple to identify a blood stain and to determine whether or not it is of human origin. If it is human blood, it may be classified into one of the four major groups "O", "A", "B", or "AB", or into subgroups based upon M, N, MN, or RH factors. However, much work has been done in trying to individualize or fingerprint blood for the identification of individuals. Although all these techniques are not yet totally foolproof, criminalistic laboratories are making good progress in this direction by using immunological techniques.

Research in hematology and serology has revealed a great number of genetic variables present in blood. It is now theoretically possible that no two individuals will have identical blood compositions. In a given population, when the probability of occurrence of an identifiable genetic variant is known and when the probability of occurrence of a combination of variants has been determined, the possible origin of a sample of blood or bloodstains containing these variants is narrowed substantially. The biochemist has used zone electrophoresis to determine the variations of hereditary factors in blood among individuals. Utilization of these methods of forensic blood analysis has been described by B. J. Cullford. Electrophoresis in general may be defined as the migration of charged ions or colloidal particles in solution under the influence of an electric field. If a solid supporting media is used to stabilize the migration into zones, the technique is called zone electrophoresis.

A recent survey of criminalistic laboratories in the United States revealed only limited use of immunological and electrophoresis analyses of blood. However, the use of the microzone electrophoresis system has been shown to provide a powerful and sensitive tool to reveal and document the individuality of blood.

P. H. Whitehead of the Biology Department of the Home Office Central Research Establishment at Aldermaston, England has recently demonstrated that it is possible to obtain a blood "antibody profile." The types of antibodies in the blood may also reveal something of the donor's health history. Immunofluorescent methods, also known as a fluorescent labeled antibody test, can now identify a number of antimicrobial antibodies, including those for diseases such as cholera, tuberculosis, thrush, vaginitis, and syphilis. A profile of the antibodies present can also reveal whether or not it is a child's blood.

A technique called the radioallergosorbent test can identify antibodies to allergies such as hay fever. Allergy tests not only discriminate between normal and sensitive persons but may also reveal where the donor of a bloodstain lived. For example, hay fever in Britain is usually associated with common cocksfoot grass. In North America it is associated with ragweed, which is rare in Britain. Thus antibodies to cocksfoot grass in the bloodstain would indicate that the donor almost certainly had lived in Britain.

Although the principal effort to date has been to direct this technique toward the identification of blood constituents, the method is equally applicable to the forensic examination of proteins in other body fluids. For example, specimens of seminal fluid, saliva, urine, cerebrospinal fluid, and tissue extracts may be analyzed in the future to characterize individuality.

The continuing interest in microzone electrophoresis indicates that in time this technique will become a standard procedure in criminalistics and in the general practice of law.

The level of trace elements in blood components such as the amount of magnesium, copper, or zinc in red or white blood cells may also be used for individual comparisons in the future.

Finally, in addition to the proper collection of evidence, analysis, and the correct interpretation of results, it is important that these three aspects of criminalistics be properly coordinated so that expert testimony can be given in a court of law. No matter how scientifically complicated an analysis might be, it must be explained and interpreted in a clear and competent manner in a court case. It is unquestionably true that one of the important satisfactions of forensic chemists is the knowledge that their work is definitely contributing to an accurate and efficient functioning of the criminal justice system.

SUGGESTED READINGS

BOOKS

Culliford, B. J. *The Examination and Typing of Blood Stains in the Crime Laboratory*. Washington, D.C.: U.S. Government Printing Office, 1974.

Feigl, Fritz. *Spot Tests in Organic Analysis.* 5th ed. New York: Elsevier Publishing Co., 1956.

Gerber, Samuel M. (ed.). *Chemistry and Crime*. Washington, D.C.: American Chemical Society, 1983.

Survey of Assessment of Blood and Bloodstains Analysis Program. Vol. I, Technical Discussion. El Segundo, Ca.:The Aerospace Corporation, 1974.

ARTICLE

Grunbaum, B. W. "Some New Approaches to the Individualization of Fresh and Dried Bloodstains." *Journal of Forensic Sciences*, 21, No. 3 (July 1976): 488-509.

CHEMISTRY AND THE PLASTIC AND GRAPHIC ARTS

Creating and Caring for Works of Art

Jonathon E. Ericson

Jonathon E. Ericson is an assistant professor in the Program in Social Ecology at the University of California at Irvine where he teaches courses on environmental health and quality. Trained at UCLA as an exploration geophysicist and anthropologist, he served from 1976 to 1978 as the conservation chemist in the Conservation Center of the Los Angeles County Museum of Art. In 1978, he joined the faculty of Harvard University and served as Founding Director of the Center for Archaeological Research and Development in the Peabody Museum at that university. Dr. Ericson is the author of over 55 articles and papers in national and international journals. He has written one book, Exchange and Production Systems in California Prehistory, *and served as co-author and co-editor of five edited volumes. His research in archaeological toxicology and biogeochemistry and other areas has taken him to China, India, Peru, Mexico and Canada as well as throughout the United States.*

Art was one of the first ways in which man expressed himself, preceding written language by many thousands of years. The creation of works of art stems from the basic needs of man to record his observations, thoughts, feelings, and religious experiences. The large cave paintings of Lascaux in central France and Altamira in southern Spain are more than 15,000 years old. Here, beautiful curvilinear forms and polychrome colors of Pleistocene animals are painted within hunting scenes. From the moment that man became an artist, he became a chemist, aware of the nature of pigments and their characteristic colors and adhesion. Concerned for the preservation of his art, he tried to choose a stable environment. In the centuries since, overlapping concerns of chemistry and art have been maintained and refined.

In modern times, chemistry has had three basic impacts on the plastic and graphic arts in providing: (1) basic materials, (2) methods for care and treatment of works of art, and (3) methods for their scientific examination.

Since the turn of the century, research has provided the basic materials which form the foundation of works of art. The artist has sought out a variety of chemical products, testing and experimenting with them. If these products worked, then they were incorporated into art. The growth of chemistry and the adoption by the artist of new materials and products has resulted, in part, in the rapid changes we see in modern art. Some claim that chemistry has completely transformed modern art in the last 20 years by contributing to a plethora of art materials and art forms.

The second impact of chemistry on the plastic and graphic arts has been to provide methods for the care and treatment of existing works of art. Chemistry has contributed to our knowledge of the deterioration of art by humidity, light, heat, unstable materials, and air pollution. More importantly, it has provided the techniques and materials for treating and stabilizing works of art. This has led to the growth of a new field called art conservation.

The modern art conservator must be well-trained in chemistry in order to examine and work with great works of art, maintaining our art heritage in good condition. Without the conservator, the visual and creative vestiges of our heritage would rapidly disappear.

The third impact of chemistry on the plastic and graphic arts has been to provide knowledge and techniques to analyze and scientifically examine works of art. Extending to prehistoric art, this research enriches the details of the historical picture that we have of the artist and his work. The scientific examination of prehistoric art is particularly exciting, because it allows us to write history which was not written before and to probe deeply into the past, unravelling its mysteries (see Chap. 3 of this volume). Fakes and forgeries are also detected by chemical methods. To meet the demand for specialized training for the scientific examination of art and prehistoric art, a new discipline called conservation chemistry has been developed.

BASIC MATERIALS

If we are going to have a clear picture of the contribution of any field, we must have a frame of reference in which to register the impact of one field upon another. Although art materials have always existed, their variety and number were limited prior to the introduction of the products developed by chemists. The history of Western painting demonstrates that the most significant changes have occurred with the introduction of a new medium or of a new way of working an older one.

Art can be classified into a number of categories — painting, drawings on paper, textiles, and 3-dimensional objects of stone, bone, wood, metals, ceramics, plastics, etc. Each category has its own history of development and set of materials.

CHANGES IN PIGMENTS

In the Metropolitan Museum of Art in New York, there is an Egyptian painter's palette dating back some 3,000 years, or to the Eighteenth Dynasty. There are eight colors in the palette: terra cotta red, light yellow ocher, medium yellow ocher, turquoise, blue, green, white, and black. These are all earth colors, ground from earth materials using ocher, carbon, copper and other minerals. The colors and materials used for pigments changed very little through the Greek and Roman eras all the way to the Age of Giotto (1267-1337) in the Early Renaissance. During the Renaissance, new pigments were discovered through simple chemical operations. As we move through time, both the number and varieties of pigments increase. Whereas Titian (ca. 1488-1576) had eight colors and white, Renoir (1841-1919) had eleven, and Cezanne (1839-1906) had eighteen. At least 250 primary pigments are now available for the painter as the direct result of modern chemical research.

PAINTING MEDIA AND THEIR CHANGES

"New art forms arise from new needs and new possibilities." As pointed out by Lawrence N. Jensen, changes in media, binder, or carrier of pigments can provide the basis for change in art form. This was aptly demonstrated at the height of the Italian Renaissance with the introduction of oil media.

There are six traditional media. The Egyptians used encaustic pigments or pigments suspended in hot wax. In fresco painting, pigments were suspended in plaster — the Sistine Chapel by Michelangelo is one of the best examples. During medieval times, egg tempera or a mixture of egg yolk and water was used in wood panel painting. Oil painting began with Hubert van Eyck (1370-1426), the Flemish master who was the first to suspend pigments in oil. Watercolor paint consists of very finely ground pigments suspended in a water solution of gum arabic. For about 200 years, pastel painting or pigmented chalk with a gum or weak binder has also been available.

Until the turn of this century, the painter was restricted to the color, texture, and form that could be created with the six traditional media. However, in 1901, Leo H. Baekeland, a chemist, discovered the first modern synthetic plastic, called Bakelite®. Since then, chemists have produced a wide array of plastics. These new materials, some of which are self-supporting, no longer limit the painter to a two-dimensional space. Painters find it hard to keep up with the evaluation of all the new products. Polyvinyl acetate, exemplified by Elmer's Glue All®, a Borden product, is one. The epoxies, like Fiberglas®, a product of Owens-Corning, is another. So are the polyesters. Also available is polyvinyl chloride resin. Paints derived from acrylic resins are perhaps the most widely accepted of these new media.

What is important here is that most of these products have distinctive and superior qualities. Each change in color, stroke, and texture produces a unique effect in the technical qualities which is, in part, the result of the material being used. The chemist plays a critical role as discoverer of art materials, while the artist plays an equally critical role as creator.

THE PLASTIC ARTS

A plastic medium is one which undergoes a transformation from a liquid to a solid or solidified amorphous state. The material can be used to model the interior of a mold while in the liquid state. Plastic media, such as molten metals or plastic resins that are liquid at room temperature, are space-filling as liquids but retain the shape of the space when in the solid form.

The beginnings of the plastic arts can be traced first to the use of clays for ceramics and then of metals in the Middle East. We know from the archaeologist that the earliest metals appear approximately at 6500 B.C. in Anatolia. At about the same time at the site of Jericho in Israel, limestone was slaked into quicklime and then poured to produce polished and decorated surfaces.

Prior to the discovery of the plastic arts, man was confined either to the graphic arts or sculptured forms produced from rocks. These media were constraining in terms of possible three-dimensional forms, surfaces, and subtle detail. From these early beginnings, the inventory of materials available increased at a constant, but slow rate. With the advent of the plastic arts, man the artist was finally free to experiment with complex forms. Too, these developments allowed him the option of reproducing particularly aesthetic or symbolic forms. Art could be reproduced, traded, and even copied by others.

The impact of modern chemistry has been to create new metals, alloys, glasses, ceramic composites, and composite materials like concrete and Portland cement. Perhaps most important for the plastic arts was the development of the "plastics" and polymer materials.

Modern or contemporary three-dimensional art can be considered a creative exploration and evaluation of new art media. The pattern is once again apparent: fundamental changes in art forms are preceded by the invention of new and usable media. New products will continue to provide the artist with materials to create and fashion new and unseen images.

OTHER ART FORMS — OTHER MATERIALS

The products made available to the art community are not solely limited to innovative developments in pigments, media, solvents, or protective coatings, though all of these are necessary for painting.

The paper industry and its chemists have produced many different materials for work with paper. A variety of paper products of different origin, material, texture, color, and acidity are now available, as well as new inks, adhesives, fixatives, and protective coatings.

Textiles, the medium of the weaver, were originally limited to natural plant and animal fibers such as cotton, linen, hemp, wool, silk, and hair. The desire to produce synthetic silk led to the development of man-made fibers. In 1884, Chardonnet, who was a student of Louis Pasteur, developed a silk-like thread. From that point on, the family tree of fibers has grown tremendously. Synthetic polymer fibers like nylon, Orlon®, and polyester are just a few of the new materials now available to the weaver or craftsman of textile art.

Concomitantly with the innovation of new fibers, chemists have developed many thousands of new synthetic dyes which are compatible with both the old and new fibers.

The development of compounds and mixing formulas for both pottery glazes and for the patina or corroded layer of metals like bronze have been the result of chemical research and the artist's experimentation. Today, glaze recipes are reported in chemical notation.

CARE AND TREATMENT OF ART

We have discussed the contributions of chemistry from a standpoint of production of art materials, which influence the art form. Whereas art is the product of creation, the art museum is the treasure house of man's expression and creativity. Within a museum, we have the responsibility for the care and preservation of works of art. Time and various agents tend to destroy our precious heritage of art representing virtually every culture known to have existed through time. The field of art conservation has grown from this need to care and restore all forms of art.

Chemistry has played, and continues to play, a critical role in the growth of this new field by providing both techniques and materials. The last decade has seen the growth of conservation centers in the larger museums and the incorporation of conservators within smaller museums. In the private sector, there are now many free-lance conservators available for consultation. As a response to the growing demand, a number of college programs designed specifically for the training of art conservators now exist.

The preservation of art is dependent on the control of the environment of the piece. Modern museums are air conditioned, maintaining their temperature between 68°-72°F. and a 50-65% relative humidity. These controls provide the conditions under which most works of art are stable. Occasionally, the relative humidity will have to be increased or decreased depending on the stability of an individual piece. Here, art conservation has borrowed techniques from physical chemistry to determine proper conditions experimentally.

The lighting of a work of art is also a critical part of its environment. Fluorescent light and sunlight contain a lot of ultraviolet light. The exposure of art to this light can cause it to fade tremendously. Papers, textiles, and organic dyes are the most sensitive to ultraviolet fading. Here, the polymer research chemist has developed special acrylic plastics like Plexiglas® UF-3, a Rohm and Haas product, which filters out the ultraviolet light.

The relative humidity, temperature and lighting are rather easy to control in a large museum. However, the agents of attack which deteriorate the appearance of the art piece are now not always so easily identified. For paintings, the accumulation of dirt and grime, discoloring of the protective varnish layer, or the tension and distortion of canvas or wood supports are destructive to the piece and distorting to us when viewing it. In the past, when cleaning a painting, people either used stronger and stronger cleaning methods like sandpaper, or, if smart, gave up. Modern chemistry has developed ways and means of using whole families of safer cleaning agents like acetone, alcohol and other organic solvents. After careful examination and evaluation, the painting conservator can remove the old varnish, reline or bond a new supporting canvas with natural wax resin or synthetic resin adhesive, spray on an isolating layer of synthetic resin, paint in pigment losses, and seal the painting with a final protective layer. All these steps use products which have been developed in the chemical laboratory.

Finally, the agents of attack are not always so subtle. Sometimes works of art fall and break, or are torn, cut, or burned.

For each case and for each object, a particular conservation method is used. Almost always, the products used or the treatment itself are direct contributions of a chemist. Without chemistry, art conservation, as we know it, would be very primitive, indeed.

SCIENTIFIC EXAMINATION AND ANALYSIS

Through the scientific examination of art, particularly prehistoric art, one is made constantly aware of man's potential as an artist-scientist. Man's ability to discover or invent particular processes and incorporate them into works of art has been demonstrated thoughout time. Each day, we learn more about how man processed his art materials and the type of signature he left during the execution of his particular work. We know more about why he chose particular types of materials and how he came upon them through wide-scale trade. It is important to note that the methods and instrumentation of an analytical chemist are used to probe into the past. We now have a clearer picture of the developments of particular technologies. Together with the archaeologist, we are defining the history of art technology, the means of processing stone, bronze, iron, glass, ceramic and many other materials into art.

With these same analytical probes of the chemist, we are working with the art historian to detail the works from the Great Masters and the present modern artist. This information makes it easier to distinguish fakes and forgeries from authentic works of art. Each day brings us more cumulative information on art and technology.

SUGGESTED READINGS

BOOKS

Baer, N.S., and L. J. Majewski. *The History of Teaching Conservation in the United States.* Venice: Triennial Meeting of the ICOM Committee for Conservation, 1975.

Birren, F. *History of Color in Painting*. New York: Reinhold Publishing Corp., 1965.

Clarke, Leslie J. *The Craftsman in Textiles*. New York: Frederick A. Praeger, 1968.

Cook, J. Gordon. *Handbook of Textile Fibers, II: Man-Made Fibers*. 4th ed. Altrincham, England: St. Ann's Press, 1968.

Crown, D.A. *The Forensic Examination of Paints and Pigments*. Springfield, Ill.: Charles C. Thomas, 1968.

Feller, R., N. Stolow, and E. Jones. *On Picture Varnishes and Their Solvents*. Rev. ed. Cleveland: Case Western Reserve University, 1971.

Gettens, R. J., and G. L. Stout. *Painting Materials, A Short Encyclopedia*. New York: Dover Publications, 1966.

Green, D. *Pottery Glazes*. New York: Watson-Guptill Publications, 1973.

Jensen, L. N. *Synthetic Painting Media*. Englewood Cliffs, N. J.: Prentice-Hall, 1964.

Plexiglas Design and Fabrication Data. Philadelphia: Rohm and Haas Company, 1973.

Robinson, S. *A History of Dyed Textiles*. Cambridge: MIT Press, 1969.

Ruhemann, H. *The Cleaning of Paintings*. New York: Frederick A. Praeger, 1968.

ARTICLES

Eklund, Jon B. "Art opens way for science." *Chemical and Engineering News* 56, No. 23 (June 5, 1978): 25-32.

Gerstner, W. "The paints of the old masters." BASF Review 2, No. 1 (January, 1977): 61-75.

Johnson, B. B., and T. Cairns. "Art Conservation: Culture Under Analysis." *Analytical Chemistry* 44, No. 1 (January 1972): 24A-36A; *ibid.*, No. 2 (February, 1972): 30A-38A.

CHEMISTRY AND HUMAN BEHAVIOR

The Next Frontier of Science

Manfred G. Reinecke

Manfred G. Reinecke, Professor of Chemistry at Texas Christian University, received his B. S. degree from the University of Wisconsin and his Ph.D. from the University of California, Berkeley. He is the author of some 50 papers on organic, natural product, and behavioral chemistry and is a member of the Chemistry of Behavior faculty at TCU. Dr. Reinecke has been the recipient of a National Science Foundation Science Faculty Fellowship and a National Academy of Sciences Exchange Scientist Award to the German Democratic Republic. A former visiting professor at the University of Tübingen and the Technical University of Wroclaw, Dr. Reinecke has also served as Chairman of the Fort Worth-Dallas Section of the American Chemical Society.

The next frontier of science is in sight. Atomic theory opened chemistry, quantum theory unlocked the secrets of physics, and the double helix theory of DNA structure led to the remarkable insights of the molecular interpretation of biology. Even as research in all these areas continues to modify, expand, and apply these key theories, the understanding of behavior looms increasingly large on the horizon as the next peak for science to scale.

This same goal has been sought by philosophers and poets, physicians and priests, politicians and police, ever since mankind has reflected on its existence. Why is one man wise and another a fool? One kind and another cruel? One good and another bad? Each of these characterizations represents a blend of behaviors which constitutes an individual's personality. Clearly, a phenomenon as complex as behavior can, and indeed must, be studied in diverse ways.

NONCHEMICAL INVESTIGATIONS OF BEHAVIOR

Physiologists have verified the belief, held since antiquity, that the organic site of behavior is the brain. Psychologists have concluded that an individual's behavior is based on heredity, experience, and environment. Much of the current research directed at the unified understanding of behavior emphasizes the molecular approach so successfully applied to biology in the past quarter-century. Whether such research is carried out by physiologists, psychologists, pharmacologists, neurologists, biochemists, or others, it is by definition chemistry, the science which deals with the behavior of matter at the molecular level.

THE CHEMICAL BASIS OF BEHAVIOR

What is the basis for the belief that an understanding of behavior must be chemical? In part this is a natural extension of the molecular interpretation of the many other processes in living organisms. For example, the molecular theory of heredity must also apply to the hereditary component of behavior. Is it not probable that the environmental component may also have a chemical explanation? This idea is supported by the long-known fact that certain chemicals can alter behavior and is reinforced by the more recent, converse observations that certain aberrant behaviors lead to chemical changes in the brain. Additional evidence comes from research on the various aspects of information processing in the brain.

THE BRAIN AND BEHAVIOR

According to current views of this process, behavior is the culmination of an intricate series of six events (Figure 1), beginning with a sensory input producing a primary electrical response in the brain. This response in turn may give rise to a secondary electrical response which generates the motor output that results in behavior. In addition, a chemical response to either the sensory or electrical activity may, in the short term, alter the electrical responses and hence motor output, and, in the long term, lead to anatomical changes resulting in more permanent alterations of behavior. Before examining the role of chemistry in understanding each of these steps, some basic tenets and tools should be mentioned.

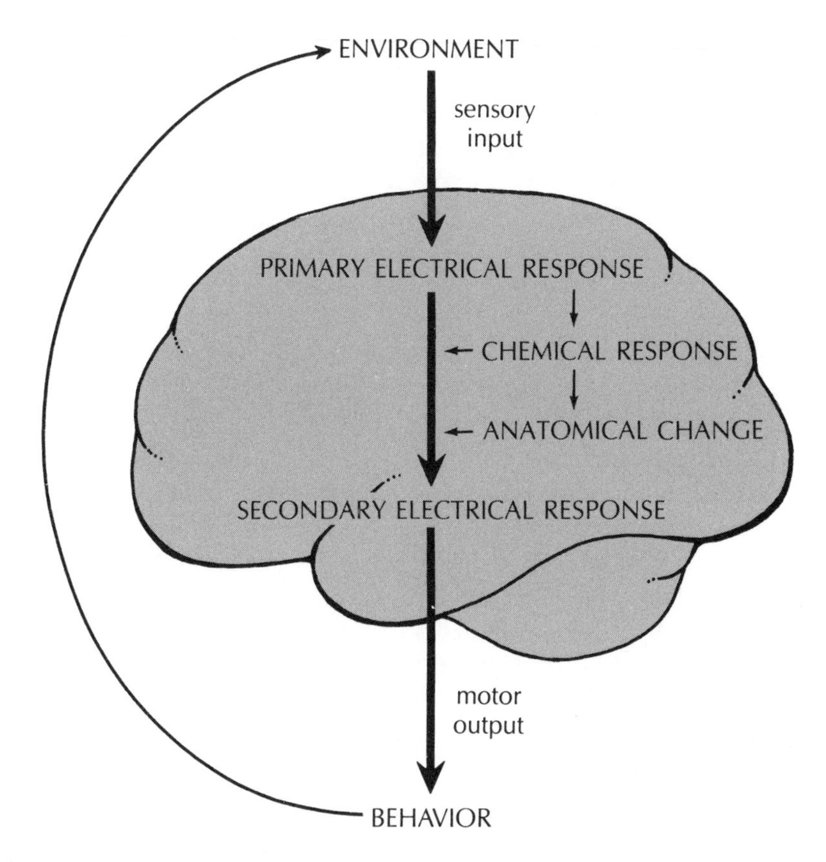

FIGURE 1: Information Processing in the Brain

Reprinted by permission from Figure 1.2, page 6 in *Functional Chemistry of the Brain* by Adrian J. Dunn and Stephen C. Bondy. Copyright 1974, Spectrum Publications, Inc., New York City.

TENETS AND TOOLS: RECIPROCITY

There are two general methods for probing the reciprocal relationship between the behavior and the chemistry of an organism. In the first, the behavior is altered and changes in chemistry are noted, while in the second, the chemistry is altered and changes in behavior are determined. Because of the complexity of the problem it is usually impossible to isolate a single behavior or a single chemical for study, and hence both approaches are usually necessary before a valid statement concerning a process is possible.

TENETS AND TOOLS: THE RECEPTOR

The central concept in chemical-biological interactions is that of the receptor. Although a few chemicals, such as general anesthetics, act by modifying the properties of biological tissues in a nonspecific way, the great majority of drugs interact with specific target sites called receptors. These sites are in themselves portions of large molecules, usually proteins which either are part of an enzyme or are embedded in a cell membrane. The chemical and the receptor have a mutual attraction for one another based on their complementary shapes and their distribution of charge and electrons. The formation of this drug-receptor complex triggers a change of some kind in the biological tissue after which the complex usually breaks apart. The receptor is thus freed to interact with an additional molecule of the chemical to produce another biological response. A very active area of research involves the isolation and chemical characterization of a whole host of receptors for different chemicals found in the central and peripheral nervous system.

NORMAL INFORMATION PROCESSING IN THE BRAIN

Step 1: Sensory Input

With this background, what then is the role of chemistry in the understanding of the six steps by which the brain processes information to produce behavior? According to Aristotle, "There is nothing in the intellect that was not first in the senses." Such sensory input arrives at the organism either as taste, smell, touch, sound, or sight.

The Chemical Senses The first two senses, taste and smell, have long been known to be chemical senses involving the interaction of chemicals with receptors on the tongue and in the nose, respectively. Considerable information on the nature of these latter receptors has been obtained from a study of the seven primary odors as a function of molecular shape. The relative importance of smell as sensory input increases among the lower animals, insects, and even bacteria, where it constitutes the dominant method of communication with the environment and with other organisms. Chemicals which are excreted by one organism and produce a specific response in another individual of the same species are known as pheromones. Many of these pheromones have been isolated, identified, and synthesized over the past two decades. They are known to trigger such diverse behaviors as alarm, attack, courtship, trail-following, territory-recognition, etc. Pheromones may be released as part of the normal life cycle of an organism or in response to specific environmental conditions. Considerable use has recently been made of sex-attractant pheromones for the control of insect populations. The role of pheromones in nonverbal communication of humans is also under investigation.

The Mechanical Senses Sound is primarily a mechanical sense. Although the same is true for touch, there is some evidence that in response to certain activating agents (toxins, bacteria, or viruses) an organism can produce a chemical known as a pyrogen which interacts with the temperature sensors (receptors) in the body to produce fever. It is also believed that the mechanism whereby acupuncture produces analgesia involves the interaction of chemicals called endorphins with the pain receptors in the body (see p.177).

The Photochemical Sense The remaining sense, sight, also involves a chemical event as its primary step. In this case the photoreceptor, a derivative of Vitamin A, is structurally altered by the light, thereby initiating a cascade of chemical and physical changes which ultimately result in an electrical signal arriving at the visual cortex.

Step 2: Primary Electrical Response

The Nerve Impulse What is the nature of the electrical response produced by the various chemical and mechanical processes which constitute these sensory inputs? A considerable body of evidence has been accumulated that these nerve impulses are due to transient changes in the permeability of the nerve cell membrane to certain electrically charged molecular species (ions). Thus, for a short period of time these ions can pass through the membrane by a process known as active transport, resulting in a net charge imbalance. The propagation of this imbalance through the nerve cells (neurons) constitutes an electric current whose duration and frequency are capable of encoding information in much the same way that a telegraph signal carries information through a wire.

The Synapse This last analogy fails in one very important way, however. Before the "message" contained in a nerve impulse reaches its ultimate behavioral goal, it must be integrated with other impulses, modified, and assigned to the appropriate motor output. This processing will involve many different neurons, which, in contrast to the telegraph wire, are not part of a continuous electrical circuit. Neurons interface with other neurons at a region called a synapse (Figure 2). The nerve impulse does not jump the synaptic cleft directly but causes the release of specific molecules called neurotransmitters from the "sending" neuron into the cleft. These chemicals are recognized by the "receiving" neuron by means of receptors. This recognition process triggers a series of chemical events which result in the continued propagation of the nerve impulse. Since each neuron is con-

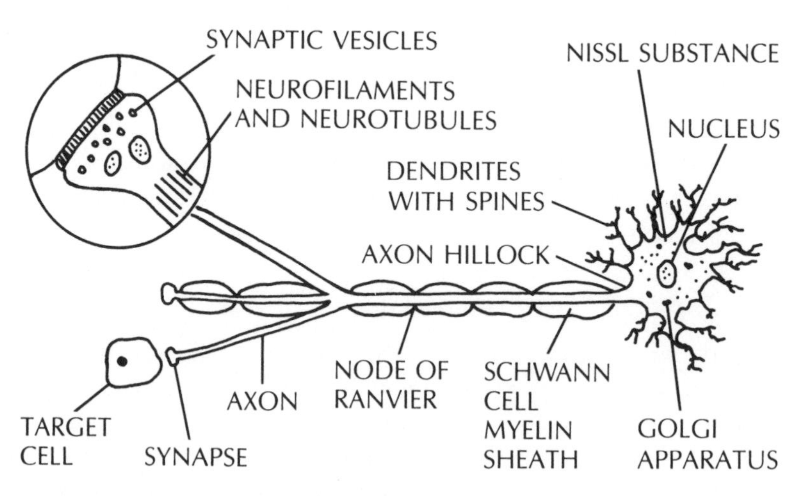

FIGURE 2: Typical Nerve Cell (Neuron) with Detail of Synapse

Reprinted by permission from Figure 6.2, page 168 in *Functional Chemistry of the Brain* by Adrian J. Dunn and Stephen C. Bondy. Copyright 1974, Spectrum Publications, Inc., New York City.

nected to many other neurons, the required "processing" of signals by dispersal, integration, and abstraction can occur.

Step 3: Secondary Electrical Response

If the original sensory input is to result in a change of behavior, the processed nerve impulse, or secondary electrical response, eventually is transmitted to a neuron which is connected to the appropriate muscle. Once again this connection is not "hard-wired" but consists of a neuromuscular junction similar to a synapse which releases a neurotransmitter that is recognized by appropriate receptors on the muscle membrane.

Step 4: Motor Output

The actual motor output is caused by muscle contractions and relaxations, the chemically based mechanisms of which are becoming increasingly understood.

Step 5: Chemical Response

Hormones In addition to the secretion of neurotransmitters at the synapse and the neuromuscular junction, the processing of sensory input can lead to other chemical responses in the brain. Hormones are chemicals released by specialized cells directly into the bloodstream which carries them to specific receptors on target cells. The hormone-receptor interaction may initiate a series of chemical reactions which lead to direct behavioral changes or secretion of other hormones. Thus hormones (and neurotransmitters) can be viewed as the vehicles of intercellular chemical communication, just as pheromones are used for interorganism chemical communication.

Neurosecretory Systems Among the hormone-secreting cells directly stimulated by nerve impulses are those in the adrenal gland, the hypothalamus, and part of the pituitary gland (Figure 3). The other part of the pituitary gland secretes hormones in response to other hormones, called releasing factors, secreted by the hypothalamus. The pituitary hormones cause further hormone release in the thyroid gland, the gonads, and the adrenal cortex. Because of the hierarchical structure of these neuro-

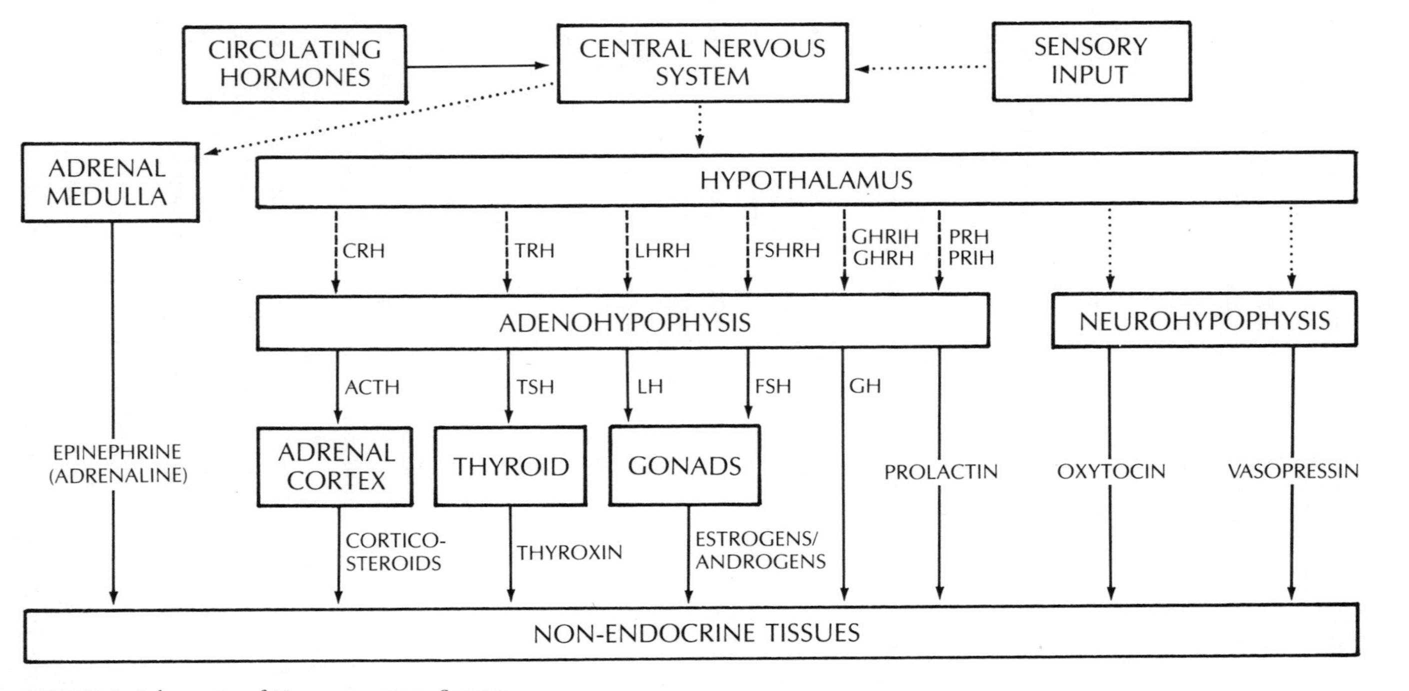

FIGURE 3: Schematic of Neurosecretory System

Reprinted by permission from Figure 8.1, page 199 in *Functional Chemistry of the Brain* by Adrian J. Dunn and Stephen C. Bondy. Copyright 1974, Spectrum Publications, Inc., New York City.

secretory systems, just a few nerve impulses may have a pronounced effect on a wide variety of functions, including growth, sexual development, metabolism, and behavior.

Neuroresponse Systems Not only can neuronal activity cause the release of chemicals, it can also respond to chemicals in the circulatory system. Feedback loops are known in which the presence of an excess of hormone shuts off any further release, presumably by some receptor-mediated process. The level of circulating chemicals other than hormones also has an effect on behavior. The presence of both sleep-inducing and excitatory substances in the cerebrospinal fluid has been demonstrated. On a longer time scale the biological clocks which control such behavior as hibernation, molting, and pupation appear to be tied to the light-dark cycles (visual sensory input) which also affect the synthesis of certain chemicals in the pineal gland of the brain. Presumably, other aspects of behavior related to the diurnal (daily) rhythm are mediated by this or similar chemical processes.

Endorphins Probably the most exciting recent discovery in the field of behavior-modifying chemicals is the detection in the brain of naturally-occurring opiates (endogenous morphines or "endorphins") which are more powerful analgesics than morphine itself. This finding promises to lead to a molecular interpretation of both the analgesic and the addictive actions of opiates and has already provided a possible molecular mechanism for analgesia by acupuncture involving the physically stimulated release of endorphins. The goal of preparing more powerful, but less addicting, analgesics related to the endorphins is of considerable medical and social significance.

Step 6: Anatomical Changes

Development The remaining aspect of information processing by the brain concerns long-term anatomical changes. It has been suggested that the level of the male sex hormone, testosterone, present in the newborn animal brain is responsible for the development of not only physiological sex characteristics but also of certain behavioral ones. The extent and complexity of sensory inputs to the newborn brain has been shown to affect directly the chemical composition and anatomical development of the mature brain and the resulting long-term behavior of the animal.

Learning and Memory The most active area of research on long-term information processing concerns learning and memory. It has been possible to distinguish a short-term and a long-term memory. From the relative ease with which they are disrupted by a variety of chemical and electrical methods, the former appears to involve transient chemical events and the latter more permanent chemical or anatomical changes. Both proteins and the RNA which controls their synthesis have been suggested as possible "carriers" of memory, but the question is far from settled. Recently a chemically-induced recovery of memory was reported in older individuals who had been given a pituitary hormone. Similar salutary effects of protein hormones on learning have been noted.

ABNORMAL INFORMATION PROCESSING

Mental Illness Since chemistry permeates virtually every aspect of normal information processing in the brain from output to input, in the short-term and in the long-term, it is also involved in abnormal processing whether caused by an internal disease state or an external chemical agent. As is so often the case with medicine, the treatment preceded the understanding of the disease. In this instance the remarkable success of the tranquilizers in treating certain psychotic states furnished a major impetus for investigation of the chemical roots of mental illness. Although such investigations resulted in much basic information and the development of many new drugs, a complete explanation is still lacking. Evidence has been presented and hypotheses formulated involving aberrations of both neurotransmitter and endorphin metabolism. Research in this area continues to thrive.

Chemical Manipulation Another major reason for the belief that mental disease has a chemical basis is the ability of many drugs to cause both similar and unique behavioral abnormalities, presumably by interacting with neurotransmitter or hormone receptor sites. Among these drugs must be included not only "hard drugs" such as amphetamines, opiates, and hallucinogens but also so-called "soft drugs" such as alcohol and nicotine. Evidence has accumulated that even the carbohydrates of a relatively normal diet can affect the levels of neurotransmitters in the brain and therefore presumably some aspects of behavior.

Clearly, chemistry already has made significant contributions to the understanding of behavior, and it promises to play an increasingly important role in the future.

ACKNOWLEDGMENT

The financial support of the TCU Research Foundation and the Robert A. Welch Foundation in the preparation and presentation of this paper is gratefully acknowledged.

SUGGESTED READINGS

BOOKS

Bloomfield, V. A., and R. E. Harrington. *Biophysical Chemistry.* San Francisco: W. H. Freeman and Co., 1975.

Calvin, M., and W. A. Pryor. *Organic Chemistry of Life.* San Francisco: W. H. Freeman and Co., 1975.

Dunn, A. J., and S. C. Bondy. *Functional Chemistry of the Brain.* Flushing, N.Y.: Spectrum Publications, Inc., 1974.

Greenough, W. T. *The Nature and Nurture of Behavior.* San Francisco: W. H. Freeman and Co., 1975.

Hanawalt, P. C., and R. H. Haynes. *The Chemical Basis of Life.* San Francisco: W. H. Freeman and Co., 1973.

Iverson, S. D., and L. L. Iverson. *Behavioral Pharmacology.* New York: Oxford University Press, 1975.

Siegel, G. J., R. W. Albers, R. Katzman, and B. W. Agranoff (eds.). *Basic Neurochemistry.* 2nd ed. Boston: Little, Brown, and Co., 1976.

ARTICLES

Bloom, F. E. "Neuropeptides." *Scientific American*, 245, No. 4 (October 1981): 148-168.

Iverson, L. L. "The Chemistry of the Brain." *Scientific American,* 241, No. 3 (September 1979): 134-149.
Rodgers, J. E. "Brain Triggers: Biochemistry and Behavior." *Science Digest*, 91 (January 1983): 60-65.
Rubenstein, E. "Diseases Caused by Impaired Communication among Cells." *Scientific American*, 242, No. 3 (March 1980): 102-121.
Wurtman, R. J. "Nutrients That Modify Brain Function." *Scientific American*, 246, No. 4 (April 1982): 50-59.

ALPHABETICAL LIST OF CONTRIBUTORS

EARLE R. CALEY
Department of Chemistry
The Ohio State University
Columbus, OH 43210

KENNETH E. COX (deceased)
Los Alamos Scientific Laboratory
University of California
Los Alamos, NM 87545

THOMAS H. DONNELLY
Department of Chemistry
Loyola University of Chicago
6525 N. Sheridan Road
Chicago, IL 60626

JONATHON E. ERICSON
Program in Social Ecology
University of California, Irvine
Irvine, CA 92717

PHILIP N. FROELICH, JR.
Department of Oceanography
Florida State University
Tallahassee, FL 32306

JAMES J. HAZDRA
Director of Health Care Education
Illinois Benedictine College
Lisle, IL 60532

NED D. HEINDEL
Department of Chemistry
Lehigh University
Bethlehem, PA 18015

GEORGE B. KAUFFMAN
Department of Chemistry
California State University, Fresno
Fresno, CA 93740

CARL PACIFICO
Management Supplements
5121 W. Penfield Road
Columbia, MD 21045

CYRIL PONNAMPERUMA
Laboratory of Chemical Evolution
University of Maryland
College Park, MD 20742

MANFRED G. REINECKE
Chemistry of Behavior Program
Department of Chemistry
Texas Christian University
Fort Worth, TX 76129

RAYMOND B. SEYMOUR
Department of Polymer Science
University of Southern Mississippi
Hattiesburg, MS 39406